INFINITE ORBITS

A BALCONY SEAT TO VIEW SCIENTIFIC DEVELOPMENTS

G S BURRA

This book is dedicated to the memory of my beloved parents.

Contents

1

Cosmic Sparklers

The Voyager II space craft hurtling away for the first ever rendezvous with the planet Neptune in 1989, suddenly received extraordinary signals from Mission Control. And its ultraviolet scanner slowly swung around.

Voyager Spacecraft(Courtesy NASA)

The International Ultraviolet Explorer, an earth orbiting satellite, and Solar Max, a sun observing satellite too got similar commands. Something dramatic had happened in February 1987, after a three hundred and eighty-three year wait. And astronomers were not going to throw the opportunity away, which may not come again in a few centuries. Apart from space crafts and satellites, sophisticated equipment was rushed to southern observatories for sighting this phenomenon. Astronomers are agog: and there's hectic activity. No wonder, at stake are answers to some of the eternal enigmas of the universe. Consider just a few of these:

The universe was born in a titanic explosion, may be fifteen billion years ago. Within a fraction of a second, three fundamental particles, the proton, neutron and electron had coalesced. Within a minute the nuclei of Helium had been synthesized. Then ensued 700,000 years of cooling during which Hydrogen and Helium, the two simplest atoms were cooked up. At that stage the universe was three fourths Hydrogen and one fourth Helium. There were no other heavy elements. How come then, that our sun contains more than its quota of the heavier elements, which in fact make up 90% of the stuff of the earth and our human bodies? Or how could simple micro-organisms through random jerks evolve into the highly complex humans?

Or, venturing a little away from home, how exactly are the mind bogging Pulsars and Black Holes born? Or, even, will the universe continue to relentlessly expand? Or finally will it collapse?

On the 23rd February, 1987, a Canadian astronomer Ian Shelton, on a stint at the Las Campanas Observatory in Chile suddenly detected a star like dot studded in the hazy Magellanic Cloud, a colony of billions of stars, visible from the southern most parts of India and further south. And that sent tele printers rattling the world over. It was probably the most momentous astronomical event in the past four centuries.

Ian Shelton

Of course the heavens had revealed such new stars before. But only, very rarely. In the past thousand years, just five had been sighted: In 1006 a few Chinese and Japanese and even Arabs, recorded that a glittering star had suddenly blazed forth in the Constellation of Lupus, low in the south. Such a dazzling spectacle was next witnessed, in 1054, in the Constellation of Taurus- again by the Chinese and Japanese. This time a few American Indians of the Anasazi tribe had spotted it too and in awe, even depicted it in rock paintings. In 1081 the Chinese and Japanese discovered yet another sparkling new star. This time in the Constellation of Cassiopeia. Then for a few centuries there was no display of these cosmic sparklers.

In 1572 however the renowned Danish astronomer, Tycho Brahe while hobbling home one night, suddenly discovered a dazzling new star, adorning Cassiopeia. Brahe studied the sparkler for sixteen long months. And with good reason. He was a staunch follower of Aristotle, the third century Greek Philosopher who had preached that the heavens were perfect – and that nothing could rupture that harmony. Why then a sudden intruding star? Brahe, a la Aristotle branded this cosmic sparkler a sub-lunary phenomenon, along with meteors. That is, it belonged to the earthly region below the Moon. So Brahe resolved to measure the distance of the new star. He failed, but nevertheless published a book, 'De Stella Nova' or 'On the New Star'. Since then the name 'Nova' caught on.

In 1604 the renowned German astronomer, Johannes Kepler detected a Nova in the Constellation of Ophiuchus. And that was one of the unluckiest misses in the history of Astronomy. Just four years later the telescope was invented by a Dutch spectacle maker, Hans Lippershey.

A supernova (Courtesy NASA)

After that something nearly happened in 1885. A German Astronomer, Ernst Hart wig detected a Nova. But that was way out in the remote Andromeda Galaxy. In fact there is some doubt whether it was visible to the naked eye at all. The fact is that it could not be fruitfully studied.

Thereafter, slowly astronomers began to pluck out hundreds of Novas with powerful telescopes. Though not many details tumbled out of these distant events, at least they salvaged a few facts and figures. For example, the distance and brightness of these objects. Moreover, helped by the growing knowledge of how stars are born and evolve, they began to understand why a star suddenly blazes for the out of the blue, literally. This is a mind-boggling story.

To appreciate it you will have to trace out the life history of stars. Visualize a huge cold, dark cloud of predominantly Hydrogen. This cloud, collapsed under its own weight, so that the interior gets heated up till the temperature at the core shoots up to a staggering 15 million degrees centigrade. At that stage thermo nuclear fusion is sparked off: That is, four atoms of Hydrogen get bashed into one atom of Helium. In the process colossal amounts of energy are vented out. This out rushing radiation stalls the collapse. A star is born. That is essentially the story of our sun, to date.

But of course after millions or even thousands of millions of years all the core Hydrogen of the star gets exhausted – and so also the out rushing energy. Then the star crumbles again. In the process the region surrounding the core gets heated up till nuclear fusion is ignited again. The star now expands and cools to become a huge reddish object. A Red Giant. The sun could become a Red Giant after a few thousand million years. By then its surface would be so close to the earth that all life would be helplessly vaporized.

Once all the Hydrogen in the shell around the core is fused into Helium, the thermo- nuclear fusion halts. There is no internal energy source now to support the Red Giant which therefore collapses again under its own weight. The core temperature in the process soars to an inconceivable hundred million degrees centigrade or so. This triggers off a new thermos nuclear reaction. Helium is bashed into heavier elements like Carbon. The out pouring energy temporarily halts further collapse. In a series of such collapses, heavy elements are cooked up in the core of a star. That is, elements right up to iron. But that's the dead end. Any further thermos nuclear reaction involving iron absorbs, not releases, energy. So the next collapse is total.

What happens then, you may wonder. For stars with a mass up to one and a half times that of the sun – and that includes the sun itself – the final collapse squeezes the atomic electrons out of their orbits. The matter then consists of a free for all mix of protons, neutrons and electrons. These particles resist further compression. It is this resistance which finally halts the collapse. The star is now a tiny, highly compressed white hot object, radiating off its remnant energy into space. How tiny and how compressed, you may ask. Imagine a sun mass, that is – more than that of 300,000 earths – squashed into the size of the earth. That is how tiny and compressed. Astronomers call such an object a White Dwarf. A human sized creature on a White Dwarf would weigh as much as a million earthlings! Interestingly, the more massive a star, the faster it swells onto a Red Giant and finally ends up as a White Dwarf.

But what happens to a star more than one and a half times as massive as the sun? Remember, it is the ultimate resistance of electrons, protons and neutrons to further compression that halts the collapse of a White Dwarf? Now, even this resistance is over shot. And, electrons plunge into protons to form neutrons. What remains is a soup of, exclusively neutrons, which resist further compression. This object, is called a 'Neutron Star'. Imagine the sun bashed into the size of a small city – just about fifteen kilometers across. And that's a Neutron star. Neutron stars not only spin rapidly, but they also behave like magnets. And they vent radio waves, light and what not from their magnetic poles. If we happen to be in this plane, we will receive the radio waves or light – but intermittently, because of the rapid spin. That is, the Neutron star emits radio or light flashes or pulses, rather like a rotating light house beacon or a rotating lawn sprinkler. Such objects were discovered and are called 'Pulsars'.

All this leads to a dramatic scenario, which arises because the vast majority of stars are born enmasse, in groups. In fact, about seventy percent of stars are 'Binaries', that is two stars swinging around each other in a cosmic waltz. Consider such a close binary pair. The more massive of the two rapidly swells into a Red Giant in a few million years.

At this stage some of its outer layer material mostly Hydrogen and Helium, spills over into the companion. While the Red Giant then collapses into a White Dwarf, the companion star which has gained mass swells into a Red Giant. In a turn table act, matter from the new Red Giant streams into the White Dwarf and gets enormously heated up in the process. This triggers off a massive thermos nuclear explosion. The brightness rapidly soars os much as a 150,000 times that of the sum. Such a stellar outburst is a 'Nova' or a new star, because a hitherto invisible binary suddenly shows up in the sky.

However, some of the Nova's observed over the past seventy-five years have been so remote, that even with this explosive brightness, they should be invisible. The blow out in these cases must be on a truly titanic scale. Such exploding "super" stars are 'Supernovas' a name coined by astronomer Fred Zwicky. The five dazzlers witnessed over the past thousand years were, in fact, Supernovas.

A similar binary system- but with a difference – can trigger off a Supernova. The White Dwarf in this case is so massive that when matter from the Red Giant rushes into it, the thermos nuclear fusion sparked off is inconceivable. Equal to a million, million, million, million Hydrogen bombs erupting in unison! Just imagine, only about a thousand such bombs erupting in unison! Just imagine only about a thousand such bombs can wipe out all civilization from a trace, while the brightness soars to a staggering 2.5 billion times that of the sun. Such Supernovas are classified as Type I. The dazzling Supernovas witnessed by Tycno and Kepler were probably of this type, as there is no remnant star left behind.

But the 1987 Supernova differs dramatically from the above scenario. Here you will have to visualize a single star, at least about ten times as massive as our sun. Very rapidly- within a few million years – through a series of contractions and expansions, the star blossoms into a colossal Red Giant with an iron core. As thermos nuclear reactions involving iron absorb energy the star now caves in on a gigantic scale within a split second. In the process the atomic electrons crash on to the protons to form neutrons and simultaneously superfast, highly penetrative particles called Neutrinos are also unleashed. So inconceivable is the collapse that the core over shoots the White Dwarf stage and is compressed into a Neutron star. The energy of the collapse itself is gobbled up by the iron to form all the heavier elements. Meanwhile the shock waves of the nearly instantaneous collapse explosively rip through the outer material, which includes Hydrogen. This is the Type II Supernova for which the brightness could mount to a billion times that of the sun. In the process all the heavy elements that have been fused are vented out and injected into the surrounding inter stellar Hydrogen-Helium clouds. Next the shock waves of the Supernova explosion, compress the new heavy element enriched interstellar cloud which begins to contract. Finally, it ends up as a second generation star like our sun, with an inherited share of heavy element content. It is rather difficult to imagine that such a Supernova blast triggered off the birth of our sun and further, furnished almost all the material of our earth and in our bodies.

This apart Supernovas might have helped fashion human beings in a far subtler manner. It is now believed that the process of evolution of one species from another was not as smooth or prosaic as Charles Darwin had originally envisaged. There were occasional mutations in the genes which were probably triggered off by increased spells of radiations or increased bursts of cosmic rays. Cosmic rays are basically protons and other charged particles which are routinely ejected by the sun. Part of the fusillade keeps colliding with the earth. But an increased burst of cosmic rays or other radiations cannot be vented by a stable star like the sun. Very likely, nearby supernovas had provided these cosmic ray spurts at different times on the history of life in earth. So, they hastened the advent of humans. Indeed, without such cosmic inputs life on earth might still have been stagnating in the oceans. But there is a chilling contra possibility: If a supernova erupts too close, say five thousand million, million kilometers away or less, its radiations could also raze out life forms on the earth- may be the very life forms fashioned earlier by other Supernovas! In fact, some years ago it was suggested that such a supernova blow out wiped out enmasse, the dinosaurs about sixty-five million years ago, a theory no longer in vogue, due to other evidence.

The 1987 Supernova is probably of Type II. At least its spectrum indicates the presence of Hydrogen which should be absent in Type I Supernovas. The first findings both comfort and confound astronomers. Thus on the 23rd of February itself four underground Neutrino detectors in Europe, America, Japan and the Soviet Union plucked out these exclusive particles- just before the explosion is that those may well be the first neutrinos to be detected from outside our solar system and the destiny of our universe hinge on these negligible particles. If they have even the slenderest of masses, the expanding universe will one day halt and collapse back on itself!

But when the Supernova has been brightening rather capriciously. Within a few days, the luminosity increased to about a hundred million times that of the sun, but slowly levelled off. After a few weeks the brightness started climbing up again. Though ever so slowly. Some puzzled astronomers have suggested that these belong to an as yet unsuspected Type III category.

For the first time in history here's a great chance to test all these theories! The Voyager II Space Craft with its ultra violet scanner and the International Ultra Violet Explorer have been sampling its ultra violet radiations. The Supernova appears to be cooling rapidly. The sun observing space craft, Solar Max is scrutinizing the supernova for its gamma rays while a Japanese satellite will be scanning it for X rays. A tantalizing question is, what will the supernova leave behind? According to theory the remnant should be a Neutron star which if suitably oriented could even show up as a Pulsar. In fact, the cloud puffed out by the 1054 Supernova is the well- known Crab Nebula and does indeed conceal a Pulsar.

But there is an even more mindboggling possibility! If the 1987 Supernova is, say about forty times as massive as our sun, the final collapse would over shoot even the Neutron star stage: The Neutrons resisting compression would capitulate! And the remnant would be a super compressed object, christened a Black Hole by physicist, J.A. Wheeler. Black Holes truly test human credulity! As you know, you need to be boosted to the right velocity to escape from any object, be it a star, a planet or a satellite. If this object shrinks, the escape velocity shoots up. Now, imagine that the sun has been crushed to the size of, not just a city as in the case of a neutron star, but a mere suburb of a city. The escape velocity on its surface would soar to the velocity of light – the ultimate speed in the universe. This means, nothing can escape from such a super compressed object. Not even light. And that's a Black Hole. Are we the privileged generation of history with a balcony seat to view the birth of a Black Hole from the 1987 Supernova?

Red Giants, White Dwarfs, Black Holes . . . sounds like a fairy tale doesn't it? Or is it a fairy tale? These are concepts on the borders of fact and fiction. Maybe fact is stranger than fiction.The coming months will reveal which way the universe will swing. As Nobel Laureate Carlo Rubbia remarked, "This is the beginning of scientific research on supernova. It was science fiction before. Now it is science fact."

2
THE SPACE OPTION

The rocket soars flawlessly through clouds of billowing smoke and the hand claps of elated viewers. But suddenly there is a catastrophic explosion. And a huge ball of raging fire plunges down through the atmosphere.

The space Shuttle (Courtesy NASA)

It had happened a few times before. For instance, the American Space Shuttle, Challenger had crashed down with the charred remains of seven instantly incinerated astronauts. But this time the fiery plummeting projectile does not spell instant doom to a billion-dollar gadget, or to a few human lives. Nothing so spectacular. However, it will kick up deadly clouds of dust which will circulate and descend over hundreds of thousands of square kilometers of the earth. And slowly seep to in millions of human beings, triggering off an excruciating and horrendous death. Not merely to these people but also to future generations born of them. For, the fiery projectile contains several tons of radioactive waste – the lethal garbage of operating nuclear reactors. The ultimate pollutant.

Next, a Space-craft with a radio- active pay load, already in earth orbit rapidly spirals down and . . ...

These two scenarios are among the horrifying possibilities one has to live with – and not just in the decades to come. In 1978 in fact, the Soviet Space Craft Cosmos 954 outfitted with a nuclear reactor spiraled in, Skylab like, and spewed radio active material over a hundred thousand square kilometer area of North Western Canada. Some of the recovered soil samples exhibited alarming levels of radio activity! Accidents in Nuclear reactors, like the Chernobyl incident, with petrifying consequences is a problem that literally just scratches the surface. A literally brewing problem lies buried below. Literally.

For, all nuclear reactors churn out a residue dubbed "Nuclear Waste", after the spent out fuel is first recycled and reusable Uranium and Plutonium are recovered. The residue is a seethingly hot brownish liquid containing mainly two hazardous radioactive elements, Strontium 90 and Cesium 137, with traces of Plutonium. The most dangerous is Strontium which burrows into the bones and bombards surrounding tissues with radiation for years on end. The highly penetrative susceptibility to lung cancer. Strontium 90 and Cesium 137 have half-lives or about thirty years. That is, they disintegrate radio activity to half their original content in thirty years. Rapid by radio activity time scales.

And yet, it would take a thousand years for the Strontium and Cesium content to dwindle to safe levels – That is, where they can be allowed to remain in the environment. So the nuclear waste had to be quarantined in underground tanks for a thousand years! Plutonium however decays into half its original content after about 25, 000 years! But because of this slow disintegration it is less hazardous.

Now consider a few spine chilling facts. Each underground storage tank contains hundreds of times the dangerous Strontium whiffs released by a single atom bomb. And there are several hundred such underground tanks, the world over. Of course, this is primarily due to all the nuclear weapons that have piled up over the decades and more recently due to nuclear reactors themselves.

Imagine a leak in one of these underground tanks! That's exactly what happened in 1975 in the U.S.A when more than 400, 000 liters of high level waste seeped into the ground, undetected and crept to within a few meters of the underground water table! A spine chilling crush with disaster!

Clearly, the question, how best to dispose off the nuclear waste? looms large over mankind. Currently the waste is usually solidified into steel canisters further cooled for a few years and then buried into geological rock formations. In fact, salt deposits are favored because, apart from indicating the absence of ground water, the rock salt deposits withstand stresses induced by the heat of radio activity, and can even self- heal any cracks formed in the process.

Geological disposal facility

Possible nuclear waste disposal underground(picture courtesy environment section of The Times)

While this is the best method with existing technology, scientists have been probing other options too. For example: Burial in holes several kilometers deep. Burial under the ocean floor. Disposal inside Arctic or Antarctic ice sheets. Two even more sophisticated options are open: First, Nuclear transmutation, that is, conversion of the waste elements into less harmful elements, but this would require an elaborate and as yet non- existent technology.

Antarctic Icesheets

And the Space Option. Why not just dump the wastes into outer space, well beyond harm's way? This could be the best, but costliest long term solution. We could blast the wastes off into a trajectory that would hurl then into the sun.

A hypothetical scenario for nulear disposal in outerspace (Anonymous Artist's Impression)

But the staggering costs would abort this scheme. Ejecting the wastes out of the solar system is an equally good option, but the economics is again not viable. Alternatively, we could blast the waste onto the Moon at a much lower cost. But could we risk the Moon which we might well colonize in the foreseeable future? Nor can we jettison the wastes into an earth orbit – The cheapest of all the space proposals. Remember how Cosmos 954 with its nuclear pay load crashed back on to the earth? Recently a mid-way proposal has emerged, which optimizes costs and safety. Why not shoot the wastes into a sun orbit, one hemmed between the orbits of the earth and the planet Venus? A Space Shuttle like spacecraft could initially hoist the wastes into an earth orbit from where it could be blasted away into a loop round the sun- an artificial nuclear micro planet. At the closest approach to the earth, the wastes would still be a safe twenty-five million kilometers away. This seems to be the most practical long term solution.

Even if we perfect nuclear waste disposal, the risk of accidents in operating reactors looms large. The hike in energy consumption in recent decades triggered off a mushrooming of nuclear reactors.More than four hundred nuclear reactors now deliver electrical energy, some of them Indian. Some Five hundred and reactors are expected to be operating, in India. Also, starting from 1979, the year of the Three Mile Island Reactor disaster in U.S.A. the number of such accidents steadily climbed up. In 1985 in fact there were 2974 reported accidents the world over. The chilling thing is the long term and excruciating nature of the consequences. For example, the Wind scale reactor in England has had an average of ten accidents per year for thirty years, including a serious fire break out which resulted in thirty- nine cancer deaths in 1983fpr example: The recent Chernobyl accident had exposed a hundred thousand people to harmful levels of radiation! They will have to be under medical checkup for rest of their lives! A few countries, recognizing the spectra of nuclear reactors are curtailing or phasing out their nuclear programs. Sweden and some other European countries for example generates about half of its electricity from nuclear reactors, but intends to shut down all plants shortly. No new contract for nuclear plants has been signed in the last years in the U.S.A. and in the last several years in England and Germany. In 1983 Spain adopted a moratorium on new plants and froze five existing ones. And so forth.

Chernobyl Disaster

While the safety of Nuclear plants is still being debated, a novel recent line of thought had emerged. Why not fling the entire problem into outer space? Nuclear reactors in space! But here too there are hazards.

First, a launch accident could send the reactor crashing down to the earth. The danger, however can be minimized in two ways: Adequate sealing and containerization perhaps in balloon like bouncing strucstures in case of a crash of the reactor, to withstand severe impacts More importantly to launch 'cold' reactors, what are non- operating ones and switch them on once they are in orbit. It is frightening to reflect that some of the Apollo Space crafts and the Voyager Missions were powered by small nuclear reactors. So are the Ulysses and Galileo Missions and many more which were to have been shot . Only, the Challenger disaster intervened and delayed. According to some experts if any of these mini nuclear pay loads had been on board the Challenger, at worst about two thousand square kilometers would have been polluted by the possible fall out resulting on four hundred cancer cases.

Next, these reactors would orbit in Nuclear safe zones. That is, in the event of the orbiting reactor showing symptoms of spiraling down like Cosmos 954, it could be switched off Immediately. And due to the high orbit, it would still rake the reactor a few hundred years to crash down. Hopefully in this period the radio activity would be toned down. Right now in fact an American test proto type nuclear reactor produced five hundred watts of electricity is afloat in such an orbit.

The Space Nuclear Reactors concept has been spurred on, unfortunately by the Star Wars Programme. The idea is to power the Star Wars space platforms with nuclear energy: Harnessing of solar energy would require huge and unwieldly solar panels that would be easy targets in the event of a space war. Ultimately millions of kilo watts of energy would be generated by these nuclear reactors to power a hypothetic calStar Wars space platforms of the bygone Reagen and hot cold war era.. In fact, the U.S.A. Planned to build a huge three hundred kilo watt reactor, SP-100 and blast it into orbit in the mid 90's. A billion-dollar project.

Out of good cometh evil out of which cometh good. That sums up the progress of Science and Technology. Rockets were the brain child of a few dreamy visionaries. But they evolved rapidly due to World War II, carefully nurtured to annihilate man and cities. And then these very rockets unveiled the space era. Similarly, a Star Wars like project, carefully nurtured to destroy could well herald an epoch of Space Nuclear Reactors. A golden epoch where much of the earth's exponentially exploding requirements of energy would be generated in the safety of outer space and beamed back via lasers of microwaves. On the other hand, Nuclear Fusion, if achieved would solve many human energy problems and right now there are many encouraging signs as we will see.

3
POWER HOUSE IN THE SKY

If you drive by any river front at sun rise you will find hundreds of half submerged, devout Hindus paying obeisance to the sun chanting nearly four-thousand-year-old hymns. This is a continuation of the thousands of years old tradition. Not only in India, but peoples of almost all ancient cultures revered the sun.

Man praying to the Sun

And rightly, we are bits of the sun, born of it and we survive because of its unflinching radiations.

Yet, till a few decades ago the source of the sun's awesome energy puzzled scientists. To give you an idea, the total energy which we humans can squeeze out of all of the earth's fuel resources, for example, coal, oil and Uranium is vented out by the sun in less than a thousandth of a second! The problem is that the sun cannot unleash such colossal amounts of energy by ordinary chemical processes without burning out within no time, within a few million years. Yet fossil records established that life existed on the earth even hundreds of millions of years ago. What then is the mysterious source of the sun's unflagging energy?

In the early part of the previous century it was realized that energy could be obtained by ripping apart heavy atoms like Uranium. This is Nuclear Fission. May be the sun was churning out energy by such fission? But soon astronomers realized that this could not be, because the sun contains negligible traces of such heavy elements.

It is easy enough to determine some of the external characteristics of the sun. For example, it is at a distance of a hundred and fifty million kilometers from us, over a million times the earth's size and is more than 300,000 times as massive. By splitting sunlight with a prism, and then analyzing it we can determine its constituents. Hydrogen the lightest atom, predominates overwhelmingly. About seventy- five percent. Another light atom, Helium constitutes about twenty percent of the sun. All other elements are present in minute traces only.

A few decades ago, Nuclear physicists were brooding over the enigma of solar energy. Clearly the key to this intriguing mystery lay buried deep in the unknown recesses of the sun. They came up with a last, desperate conjecture. Suppose light elements like Hydrogen could be bashed into heavier elements, the exact reverse of fission? May be energy is released in the process? It turned out that this is possible! So the riddle of solar energy led to the discovery of Nuclear Fusion, which may well ensure man's future survival!

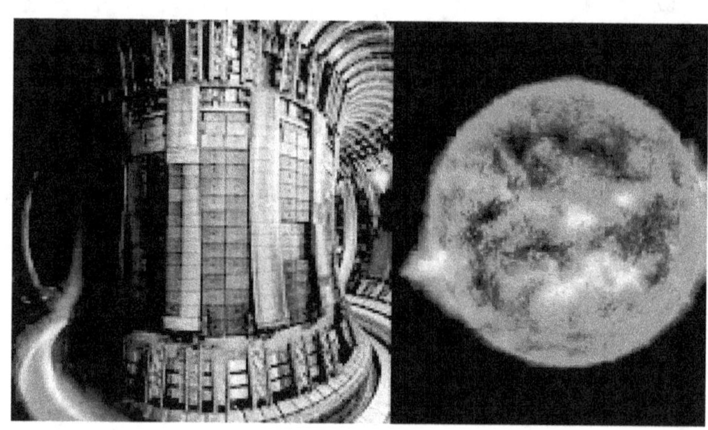

Nuclear reactor and the sun (Courtesy NASA)

In the deep interior of the sun, four atoms of Hydrogen get bashed into one atom of Helium, at a temperature of about fifteen million degrees Centigrade. This disgorges the sun's awesome energy. In the process the sun is losing mass at the rate of four million tons per second! This energy rushes out in the form of soaring X rays which seep through 500,000 kilometers of overlying solar material. But how! streaking along with the speed of light, you would expect these X rays to slice through the sun within seconds. But curiously enough these high energy rays weave through and bounce off the atoms, rather like a drunk. They zig zag and reach the sun's surface. In the process they are toned down into relatively harmless ultra violet radiation and light.

The smooth, unchanging and blindingly bright surface of the sun springs a big surprise! It is in a state if ceaseless, violent activity. Careful examination reveals huge tongues of flaming gas that keep climbing to a height of about 10,000 kilometers and then crash back into the sun -- All within about 10 minutes. At any instant about 10,000 such turbulent columns of gas are looping around. Less frequently huge streamers of gas or prominences soar to a height of 200,000 kilometers and plunge back on to the sun, a million kilometers away.

All these are routine features of the quiet sun! Occasionally there are outbursts which defy imagination: The solar flares, which are linked with the dark and transient sun spots (see, "When the sun changes its spots"). Flares are colossal gas columns unleashed within just about thirty seconds. They are triggered off when the sun flushes out energy equivalent to ravaging floods of charged particles like Protons and Electrons are disgorged. Such particle streams are called solar winds. Part of this stream collides with the earth, a hundred and fifty million kilometers away, at speeds of up to six hundred kilometers per second. Luckily for us, the earth is swaddled by a magnetic field, which deflects this tide of particles. Even so they disrupt telecommunication, and occasionally power supply in the higher latitudes. In fact, by such streams of particles. One of the strangest reported consequences is that those intense solar winds triggered off a series of magnetic naval mines during wars!

The turbulent sun which is a cosmic nuclear reactor seething away in the emptiness of Interstellar space, becomes a spell binding spectacle during a total solar eclipse. By a quirk of the cosmos, the Moon appears almost exactly as big as the sun. So it can just block the visible part of the sun during the eclipse. That's when the sun's invisible atmosphere comes alive! A gorgeous crown like halo, called the Corona. It extends up to millions of kilometers away from the sun, and its temperature climbs up to almost two million degrees centigrade. The Corona has enthralled people for eons. But only in the early forties, scientists realized that it is a thin atmosphere girdling the sun, made up of the same material. In fact, the Corona is almost all vacuum. Better than the best vacuum that can be created on the earth!

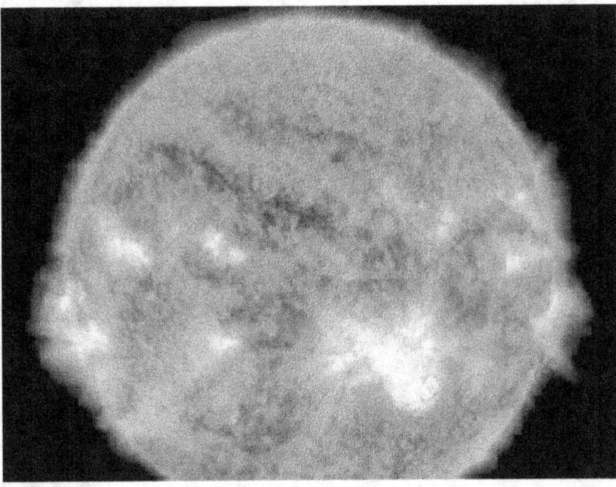

The Sun Courtesy (NASA)

In recent years the sun has been observed by several space crafts which have tuned into some of its radiations that are otherwise blocked out by the earth's atmosphere. At least fourteen exclusive Solar Observation Satellites have been floated by NASA and the European Space Agency. For example, Solar Observatory aboard the American Space Station Skylab, 7 orbiting Solar Observatories (OSO's), three International Sun Earth Explorers (ISEE's) the Helios Satellites and the Solar Max Mission (Solar Max) and so on. All these examined the sun in the X ray, ultra violet and Infra-Red regions. Pockets of intense X ray activity have been detected. These seem to be associated with solar flares. Minor fluctuations in the solar energy output have been recorded. Mysterious 'Holes', that is Black regions have been discovered near the sun's polar regions and in its Corona. Streams of charged particles, the solar winds, are ejected through these coronal holes. The Corona itself extends deep onto the solar system in the garb of solar winds. In this sense we are immersed in it. In 1983 U.S. Space Craft Pioneer 10 hurtling out of the solar system detected solar winds in the vicinity of the outer most planets.

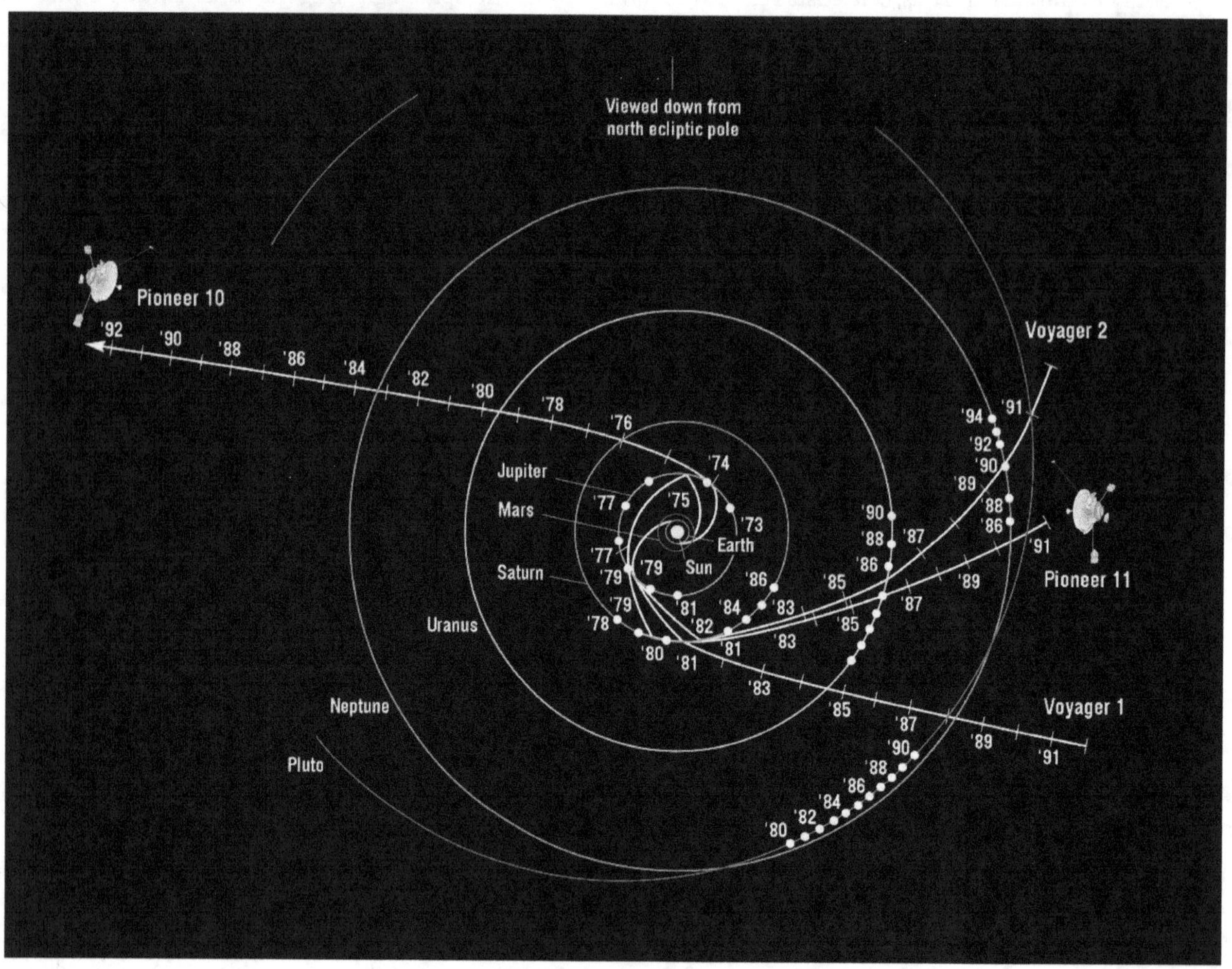

Trajectory of the Pioneer Spacecraft(Picture courtesy NASA)

Inspite of all these Space based discoveries, we are as yet in the dark about how exactly sun spots and associated flares, prominences and solar winds affect the earth. Apart from communication break-downs and induced disruptive voltages, solar winds and sun sports seem to be linked to weather patterns and even agriculturally produces on the earth! Today astronomers are eagerly awaiting the development of "Solar Meteorology". This would be a boon indeed. Whether it be for agricultural planning on the earth or for protective action against astronauts being exposed to dangerous doses of solar radiation.

The sun exhibits other idiosyncrasies. It keeps pulsating, just a wee bit. By about ten kilometers every hour. Probably energy imbalances within the sun cause this flexing. Another enigma has surfaced in recent years. The Nuclear Reactions deep inside the sun unleash high speed particles called Neutrinos. These can easily slice through the sun. The earth and other energy particles too. In fact, Neutrinos can penetrate right through millions of millions of kilometers thick load slabs! However, they can be tracked down by rather strange footprints that they leave behind. When a Neutrino collides with an isotope of chlorine called Chlorine- 37, another Isotope, this time of Argon, Argon 37 is produced. Experiments with Chlorine 37 stored in tanks deep below the earth's surface in a gold mine in South Dakota, U.S.A., and also other places, have thrown up a perplexing result. Only about a third of the expected number of Neutrinos showed up. Where is the discrepancy? In our understanding of the sun or of the Neutrinos? These days detectors below the antarctic ice have been used to detect anomalies like the Glashow resonance can as we will see later.

Over its almost 5000-million-year history, the sun has devoured half of its Hydrogen fuel. But there is enough still to keep the sun seething for another 5000 million years. After about 3000 million years however, the sun would swell and its surface temperature would drop. That is, it would redden and blossom into a Red Giant Star. With such a low surface temperature, the earth would normally have frozen into a chilling life- less world. But because by then the sun would have bloated beyond the orbit of the planet Venus, it would be so bright on the earth, that paradoxically, the oceans would boil off. The earth itself, with all its marvelous life forms and its priceless secrets of beautiful bygone civilizations would evaporate off in the heat of this light.

4
GRISLY NIGHT OF THE COSMIC HAILSTORM

It is curious, but the presence of nothing, poses what has been, dubbed, 'the greatest of all titillating puzzles'. One which has let loose an avalanche of mind- blowing theories, some ominous.

But first, you will have to excavate wide tracts of the earth. Initially you will scoop out sedimentary deposits jammed with the bones of mammalian ancestors. Elephants, horses and the likes. Next as you delve deeper into the earth and its past, you will shovel up nothing, except of course, rocks. Meters and meters of them.

And just as you give up of sheer ennui.... A stupendous discovery! You have gouged up a ghastly mass cemetery! Bizarre bones. Any number of them. Nothing quite like anything conceivable.

For a hundred and fifty years, scientists have meticulously assembled these bones together. And they have fashioned an awesome scenario for you to view. One dominated by frightful mammoths!

Columbian mammoth in the Page Museum in Los Angeles.

The epoch: About two hundred and Seventy million ago. All of the continents are fused together in one sprawling land mass. The earth is a vast swamp. Sub- tropical jungles abound. The most imposing inhabitants, lording over these grotesquely beautiful stretches are the titanic dinosaurs or 'Terrible Lizards'. Here, over three hundred bewilderingly diverse types of dinosaur's tussle for survival alongside a host of other prehistoric beasts.

The twenty- five meter, eighty tone Ultrasaurus. A vegetarian colossus with a twelve-meter crane- like neck. Probably the largest creature ever to trod on the earth. Or, the almost equally huge Diplodocus, also herbivorous, swiping its long, powerful tail – a deadly weapon. These titans were so massive, they would sink into the ground. Luckily, they remained immersed in the swamps, where the water buoyed them. Some of the dinosaurs had no teeth. They gulped large stones which burrowed in their stomachs and ground the food consumed. Others bore two hideous rows of hundreds of terrifying teeth. To rip flesh or crush bones or simply to chew wood.

Artist's impression of dinosaurs (courtesy New Zealand geographic society)

There you could glimpse a gigantic dinosaur lumbering out of the Swamp. On to drier land. The slightly smaller, but the deadliest of carnivores, the allosarus, is lurking for just this opportunity. He heaves himself up on his powerful hind legs and lunges at the ultrasaurus, Kangaroo like. A huge spiked tail rushes through the air and crashes into him. He totters. But only momentarily. He lurches again towards his prey. Jaws, claws and all. The most horrific battle the earth has beheld!

And here you would confront another encounter of naked terror. A squadron of terrible, flying dinosaurs, the pterosaurs, swoop on to helpless victims- mammals, or possibly, other dinosaurs. Renting them apart, soaring away. Shrieks. Screeches, gushing cold blood. . .

For a hundred and fifty million years these monstrous marvels, fashioned by nature over millions of years of experimental evolution, reigned the earth as undisputed overloads. One of the most enduring success stories, ever. Grotesque witnesses to today's continents wrenching apart and drifting away at an excruciating crawl, to their present positions. Over millions of years.

And suddenly, these terrifying titans vanished! About sixty- five million years ago, drama and trauma struck. Wiping out the whole species ruthlessly.

The domineering dinosaurs vanished within an eye blink. Had they waged a losing battle of survival and leisurely faded out? you wouldn't encounter the steep underground stretch, barren of fossilized bones. Rather, the dinosaur fossils would progressively dwindle as you approached the earth's surface from below.

One of the most enduring enigmas of modern science is, what was that terrible event? This riddle has hustled out hypotheses which are bewildering for their diversity and breathtaking for creativity.

Most of the theories flounder because they model a phased and partial decline. Not abrupt, total annihilation. Theories like epidemic diseases or evolutionary senility. Speculations about rising mountains, receding sea levels or drifting continents, fare no better.

So, frustrated scientists have hazarded a way out of conjectures. For instance, a species of flowers, angiosperms, blossomed profusely, just about the time the dinosaurs had a long trip with no return.

Consider two other premises. Small mammals swarmed the earth and preyed on the dinosaurs' leathery eggs on a devastatingly vast scale. Or, global temperatures suddenly soared, or even swooped. And heat-or-freeze sterilized the cold-blooded monsters' sans mammalian thermostats. Sort of a mass vasectomy.

And then, the cosmic twist. Thwarted scientists swung to the view that such a sweeping holocaust could only be let loose by extra-terrestrial inputs. A whole new breed of theories has surfaced. Exotic extinctions!

Like, a supernova exploded, millions of millions of kilometers away. An equivalent of a million million million Hydrogen bombs erupting in unison. Lethal radiation was spewed all over. Just a wisp of it engulfed the earth, razing it. Blitzing the dinosaurs and what not. The very supernova which triggered off the formation of the sun, the very exploring in which atoms that are in the streams of creating like ourselves. Yes, such mega bursts which start life, could even annihilate it.

Another conjecture is even more intriguing. The earth's magnetic field keeps afloat a vast belt of charged particles from a few hundred kilometers upwards. The Van Allen radiation belts. These shut out lethal radiation, like the ultraviolet rays, which are continuously ejected by the sun. But there's a curious mystery that has tantalized scientists: The North and South magnetic poles of the earth keep switching – with varying periods of over ten million years. In between, the earth's magnetism dwindles to almost nothing. All this, by the way, implants screaming clues in magnetized rocks that are vented by volcanoes, or which cool their heels on ocean beds. This periodic flip flop of geomagnetism distorts the Van Allen radiation safety belts. And every once a while, deadly ultraviolet and other waves gatecrash, blazing and razing across the earth. What hope for the hapless titanosaurs?

Then, as an astonishing 1979 discovery unleashed a series of mind boggling hypotheses.

Father and son geologists, Luis and Walter Alvarez of the University of California, detected extraordinary doses of the rare metal, iridium in clay samples of the Dinosaur Death epoch. Stray meteorites from outer space could transport the iridium. But, the Alverezs reckoned that only a global, phenomenal event could account for its fairly sprawling distribution. Like a huge object hurtling from the cosmos and smashing into the earth. And, to bolster this conjecture, such high iridium doses soon surfaced from even remoter regions!

Review the scenario: A huge boulder, an asteroid in fact, over ten kilometers across, would swing so close to the earth- well it would plunge down and ram into it. Tantamount to millions of hydrogen bombs blasting the earth. The super powers nuclear arsenal, thousands of times over!

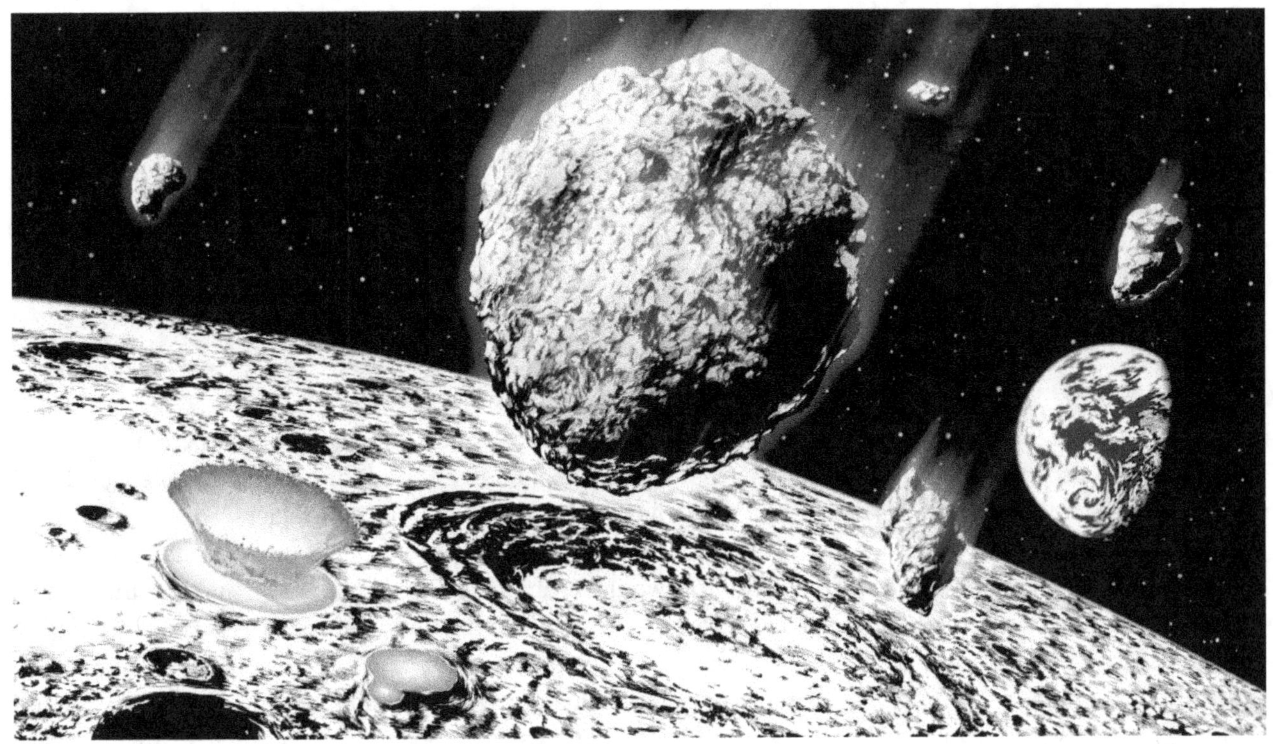

Artist's conception of asteroids(Image courtesy Osaka University)

Imagine the pillars of dust kicked up! Trillions of tones! Which would swaddle the earth, blanket like, blockading vital heat and sunlight. Dust thrown up in volcanic eruptions too, is notorious for this, though on a micro scale. Photo synthesis would dip. Temperatures would swoop. Plants would wither. And whole species like the dinosaurs would freeze- starve to death.

Later, three American scientists, E. M. Shoemaker, C. Emiliani and E. B. Kraus have twirled a half twist into the theory. A gatecrashing asteroid would more likely, splash into an ocean. Towering fountains of water would whoosh up, strangling the atmosphere in a vapor blanket. With reverse consequences. Heat would be trapped in the atmosphere. Turning the earth into a sort of gargantuan greenhouse inferno. The temperature would soar globally. And . . . You can imagine the rest.......

5

TEEMING WORLDS

Suddenly evidence is flooding in which corroborates an ingenious conclusion that astronomers have veered to, over a few patient decades: The thrilling prospect that the earth is but one of millions of inhabitable worlds seems guaranteed. Indeed, today spacecrafts like Kepler which have sent back several positive feelers.

Scientists do concede that life as we conceive of it, can survive within bewilderingly wide limits of temperature, radiation, atmospheric content and so forth. But all this would be characteristic of planets swinging round stars. Not the stars themselves, which are seething nuclear infernos. So the number of inhabitable worlds is directly linked to the number of stars hanging on to planets.

Unluckily, even today's super optical telescopes are too feeble to dig out possible beyond our solar system. Not even for the nearest star: The planets just vanish in the star glare. Astronomers are perfecting instruments that will obscure the star glow and reveal the planets. By today we could actually be glimpsing these exotic worlds. The Space Telescope, will hopefully has yielded precious clues. Even definite answers.

Till very recently however, astronomers relied on two rather indirect, if ingenious, leads:

A Star, as it rushes through space – and all stars are – would rock and wobble due to the tugs of planets in tow. Much like a harassed mother on a stroll, being wrenched by her children. All this shows up as wiggles and jiggles on the path of the star as it crawls across the backdrop of the constellations. A few stars have indeed displayed such predictable wobbles. The catch is, this could also be set off by companion stars. Not necessarily planets.

A good example is Bernard's star. Visible only through telescopes, less than six light years away, this Red Giant is our second closest star.

The 48-inch Samuel Oschin Telescope at Palomar Observatory (Image credit: Caltech.edu)

Just beyond the alpha Centauri system. Its trajectory has been scrupulously scrutinized for over five decades. Astronomers conclude that it may well be girdled by two planets: A Jupiter sized giant orbiting the star every eleven and a half years and a more distant companion, half as massive, whirling round once every twenty years or thereabouts. The nearby star, Epsilon Eridani too, exhibits similar jiggles. It is conjectured to be another planetary system.

The other evidence favoring planets is more theoretical. It pivots on the theory of the solar system's origin. If a freak accident brought forth the planets, they must be scarce. But suppose planets are churned out by a perfectly natural process. Then the universe would abound with planetary systems. For a few centuries theories swung between either extreme. From planets spilling out of the solar material after a breathtaking collision between the sun and a comet or a star, to the steady evolution of planets from rotating material surrounding the sun. The destiny of inhabitable worlds and ultimately of life itself, in the universe, hinged on this pulsating debate which is today of historical value. Over the past few decades, astronomers have finally blueprinted the following scenario.

Visualize a huge star, embedded in a colossal, if thin cloud or Hydrogen. About 4,500 million years ago, the star suddenly erupted. A million million million million Hydrogen bombs exploding in unison would pale before such a cosmic sparkler. It's a commonplace at that Shock waves from the inconceivable blow out shattered the cloud into shreds.

Now follow the fascinating history of one of these tidbits. It is billions of kilometers across, though. It spins and shrinks and spins faster. And shrinks further. And so forth. Till it rotates so furiously that huge chunks of matter are spewed out by centrifugal forces. Rather like bits of mud sticking to a wheel, being ejected as the wheel spins rapidly. The vented material settles into a sort of a disc enclosing the original cloud that has now graduated into the sun. The particles of the disc, through eons of collisions, clump together. Finally, these clumps evolve into the nine known planets and the asteroid hordes.

This theory accounts for many of the finer details of the solar system. There is some evidence too, for the initial stellar blowout that triggered it all off: The planets, reptiles, the Taj Mahal, slums, nuclear wars . . . You see, the oldest meteorites which pound the earth even today, condensed out of the antique planetary disc. It is striking that they are about 4,500 million years old – the age of the solar system itself. Even more remarkable, they exhibit the decay debris of a very radioactive isotope of Aluminum, Al- 26, now extinct. This disintegrates so rapidly that the meteorite and the isotope must have formed at about the same time. And, this Aluminum isotope can only be concocted in such a stellar eruption.

More recently, theoretical computer simulations with clouds of gas and dust – the foetus of the stars – has thrown up a fascinating spectrum of star systems. From a star besieged by a swarm of asteroids through full-fledged planetary systems down to star pairs swirling about each other.

So astronomers are convinced that planets are a cosmic commonplace. Here's how astronomer Fred Hoyle opens a chapter in his book, Frontiers of Astronomy: ". . . In a former discussion of this particular problem, stars of the Milky Way. . . The new number comes out, not at a mere one million, but at 100,000 million..."!!

And now there's a deluge of hard core evidence for the existence of planetary worlds. The Anglo-U.S.-Dutch Infra-Red Astronomical Satellite, IRAS, triggered it all off. Launched on twenty- fifth January, 1983, IRAS scanned the heavens with a super cooled infra-red telescope. Soon it sentback a sensational discovery. Twenty-six light years away, the brilliant star Vega(Abhijit) seemed to be clinging on to a disc of asteroids, flung up to about twice Pluto's distance from the sun. Would be planets? Astronomers certainly think so. And, interestingly, Vega is younger than the sun. Later in 1965, IRAS unearthed another planetary system in gestation. This time around the remote star Fomalhaut.

Next, in 1984, IRAS reported that the star Beta Pectoris, fifty light years off, puffed abnormal amounts of infrared radiation. This suggested that solid material whirled round the star. Sure enough, a team from the University of Arizona and the Jet Propulsion Laboratory in California, employing special computer aided techniques, snapped up a vast swarm of solid particles and boulders swirling round the star, spanning over a hundred million kilometers. And probably with the same composition as our planets, too. Yet another preview of planets being charmed out.

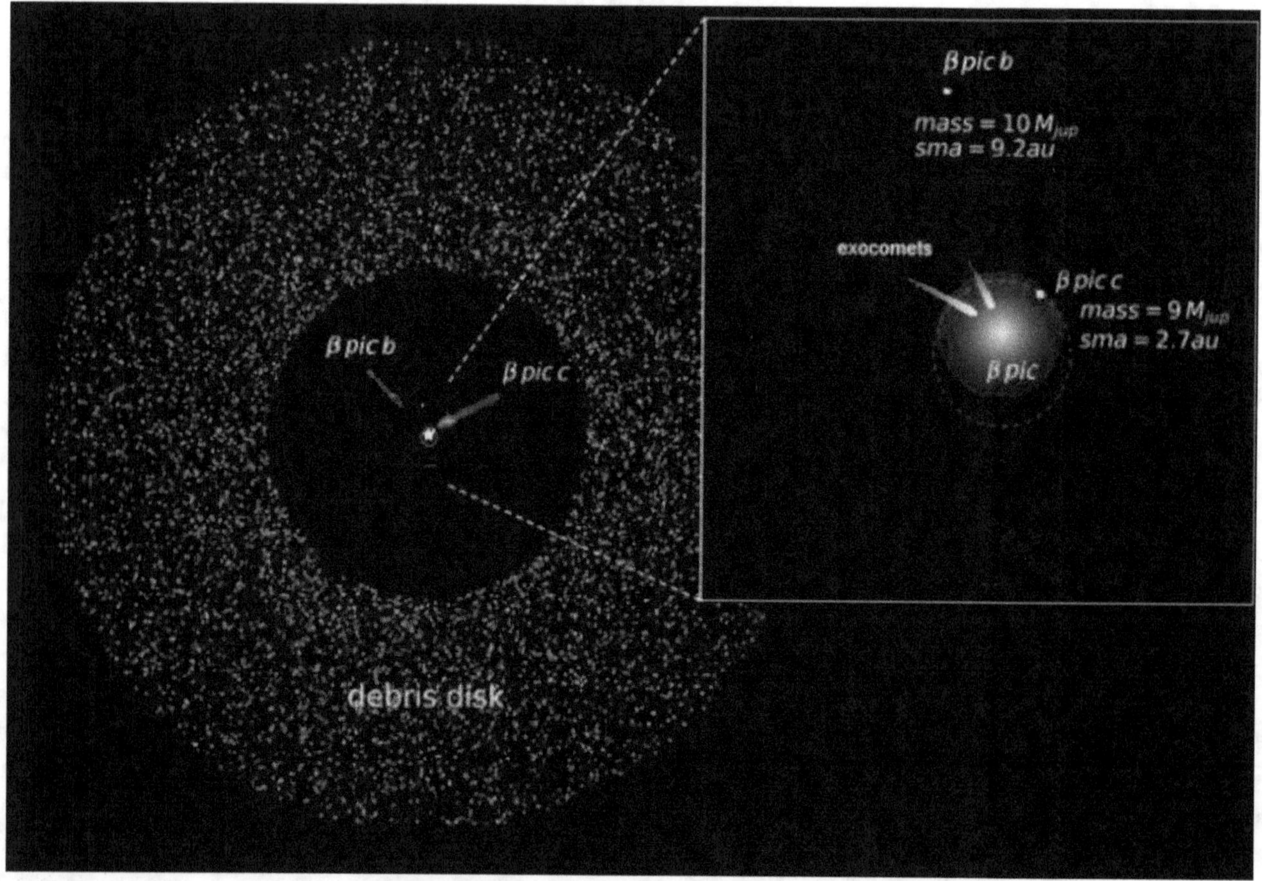

The Beta Pictoris planetary system. Image credit: P. Rubini / A.-M. Lagrange.

In mid-1984, the inconspicuous star Van Biesbroeck 8, twenty-one light years away, divulged an intriguing secret to a team from the University of Arizona and the National Optical Astronomical Observatory of the U.S.A., which employed a computerized infrared technique: A mysterious companion, a shade smaller than Jupiter, very much hotter and at least fifty times as massive, lurked a thousand million kilometers away. This could well be a planet. But there is another startling prospect. It could be one of a to- be discovered species of stunted stars, christened 'brown dwarfs'. These do not have enough matter to spark off their internal nuclear reactors. Too faint to be detected these conjectured curiosities, a cross between stars and planets, were theorized by India born astrophysicist S.S. Kumar of Virgins.

Using data from ESO's High Accuracy Radial Velocity Planet Searcher (HARPS) instrument, a multinational team of astronomers has uncovered evidence for a second giant exoplanet orbiting in the nearby, young star Beta Pictoris (β Pictoris).Two giant planets, Beta Pictoris b and c, orbit the nearby, young star Beta Pictoris. Image credit: P. Rubini / A.-M. Lagrange.

By the way, if the new find is indeed a brown dwarf, the consequences would be dramatic. There would be so many of these 'pulsars' if you want, that under their extra wright the expanding universe would ultimately collapse and vanish into oblivion!

In case you are a diehard skeptic, here's one more pointer for good measure. A team of American, Soviet, German and Swedish radio astronomers, scanning with computer aided radio telescopes from different locales, detected a sun-like star-in-the-making, in the Great Orion Nebula. This proto-star is hugged by . . . You've guessed it – the typical pre- planetary disc, flung to almost a thousand million kilometers.

The outburst of evidence for planetary systems in our closest vicinity itself, does indeed bolster the belief that star spawn planets prolifically. Even if a tiny fraction of the stars churns out planets, they would be swarming in the cosmic swamp, bacteria like.And then came the Kepler and so many others. There is a famous equation due Drake that endorses this.

And, given an average planet, scientists are convinced that sooner or later some form of life would form there. . .
But that is another story.

6

TRIUMPH AND TRAUMA

The Halley show climbed to a crescendo when an earth armada of 5 Space crafts skirmished the comet. Those exciting encounters throw up surprises and intriguing die- crafts, without actually toppling current Comet theories.

Two Soviet Vega space crafts looped past the Planet Venus, parachuting instrument packages on to it, before hurtling towards Halley's Comet. On March 6th Vega I observed to within 8500 kms of the comet's nucleus at an awesome speed of 200,000 kms per hour. Over a 3-hour span, it beamed back 500 views of the comet. About half of the space craft's solar panels were destroyed even though it seemed to rom into less than the budgeted barrage of dust. It hinted that the comet's icy nucleus, exposed to the soaring solar radiation was evaporating rapidly. Infect twice or even thrice as rapidly as expected. Unconfirmed first reports suggested that the Vega craft might have Just plucked out traces of hydrocarbons in the comet. This would be a spectacular discovery, if confirmed. It would bolster a very recent, and intriguing, though unconvincing theory proposed by celebrity Astronomer Sir Fred Hoyle and Dr. Chandra Wickrama Singe, that comets transported from deep space, the first germ of life to the earth. Or, they may even now be unleashing viruses of a number of diseases on to unsuspecting humans! Vega I helped chart out the comet's exact course too. So, pilot van like, it enabled last minute corrections to be executed in the trajectories of the Vega II and Giotto space crafts.

These engineering models of the Vega landers' spacecraft body and landing apparatus are housed at the Steven F. Udvar-Hazy Center in Virginia, part of the Smithsonian's National Air and Space Museum.Photograph courtesy of Smithsonian's National Air and Space Museum

The Japanese Suisei space craft skirted Halley's Comet the 8th of March. It scanned the comet with an ultraviolet on camera from a distance of 150,000 kms. The 10- million-kilometer-wide coma of the comet appeared to brighten and darken periodically. As if the comet blinked every 53 hours. This fits in well with an earlier observation that the comet's nucleus spins every 50 hours.

On the 9th of March, Vega II skirmished the comet, flashing back some 700 glimpses from barely 8000 kms away. Swarming dust clouds knocked out half of the space craft's solar panels and two of its experiments. An enigma was that Vega II encountered much less dust than had Vega I barely 3 days earlier. Why had the dust level suddenly swooped? However, the Vegas confirmed that the nucleus of the comet is indeed solid and ferreted out the expected presence of magnetism.

Venus. Credit: NASA/ROGER RESSMEYER/CORBIS/VCG

Saki Grake the second Japanese probe streaked past the comet on the 10th of March from a distance of about 7 million kms. It probed the solar winds engulfing the comet.

The piece de resistance was the European Space Agency's Giotto space craft. In a suicidal encounter on the 14th of March, the space craft, hurtling at the dizzy speed of 250,000 kms per hour, rammed into the densest region of the comet. To within a breathtaking 550 kms of its nucleus. It relayed back 2000 images of the comet. And of course streams of other data. But just 2 seconds before the closest brush, Giotto blacked out. Apparently it had plunged into a dust wall that had swung the space craft's antennae away. Though its camera had been destroyed, Blotto won back again transmitting weak signals after about half an hour. A real bonanza. It looks like the comet's nucleus is very irregular and elongated. Rather like an unopened peanut. or a Potato. Just about 3 kilometers by 15 kilometers. About twice what was originally suspected. A surprise is, that the nucleus is dark. One of the darkest object in the solar system, in fact. Possibly the lay nucleus is peppered with densely packed, embedded, block particles.

Though astronomers will take months to part out the flood of data, the prevalent theory seems to hold ground! comets are most likely, chunks of dirty loo ejected from a huge belt of the primitive material of the solar system, girdling the sun.

The United States to the chagrin of its scientific community has taken a back seat, away from the comet spotlight. Budgetary squeezes forced NASA to scrap an ambitious Halley cum comet Encke intercept mission. Instead, as a sop, an earth orbiting space craft, IC 111, launched to study solar winds was roped in. Originally, to be deflected for a brush with Halley. But the space craft's signals turned out to be too feeble to be picked up from the vicinity of the distant intruder. So, as a further climb down, it was diverted to encounter the tiny, nearby comet, Giacobini - Zinner which swings by every six and a half years only. In September 1985, for the first time in history the space craft ploughed through the tail of the comet for about 15 minutes. It confirmed earlier spectroscopic analyses and weeded out traces of carbon monoxide, water vapor and dust, in the comet.

But somber American space scientists had a sudden lucky break with comet Halley. Just when the comet swooped closest to the sun, in early February, it swelled to a colossal size. Its coma measured 20 million kms across, a spectacle not visible from the earth.

Stunning image of Halley's Comet, obtained in 1986: Credit ESO

However, on observer on the Planet Venus would have ogled all of it. And what better observer then the space craft Pioneer 12 which had been orbiting the planet Venus December 1978 on a surveillance mission? Promptly opportunistic north based scientists at the Amen Research Centre in California swung Pioneer Twelve's ultra violet scanner towards Halley. The space-craft observed the comet till March 6th, discerning a huge loss of water due to evaporation. Up to 70 tons per second. In fact, the comet's nucleus seems to be shedding about 9 meters of its material on each trip!

It was to be NASA's busiest year in Space. The schedule included 15 launches of the 4 Space Shuttles. The budgeted turnover in terms of communications, satellites and Commercial experiments mounted to billions of dollars. Not to mention a few breathtaking projects.

All that blew up in the horrendous explosion of the Space Shuttle Challenger, on the 28th of January. With 7 Astronauts aboard, it was the worst ever Space disaster. Also blasted to shreds was a hundred million dollar NASA Satellite. And an experiment to sample the ultraviolet spectrum of Halley's Comet.

The shuttle missions in the coming months included, apart from the floating of military and communication satellites and a Halley Observation in March, At least 3 spectacular much awaited projects: In May, Challenger was to unleash a NASA-European Space Agency space craft, Ullysses, towards Jupiter. After hurtling towards Jupiter Ullysses would be catapulted into on orbit over the polar regions of the Sun, Man's first ever glimpse of these Solar poles, Infect. In Ray itself the shuttle Atlantic would have sent the eagerly awaited space craft, Galileo, swinging into an orbit around Jupiter. Galileo would then release probes into the atmosphere of this an awesome planet. And, in October, Atlantis again, was to float the overdue dollars 1.2 billion Space Telescope. The eye in the sky to end all eyes.

But NASA had frozen further shuttle missions. By past form the shuttles could be grounded for at least a year. In January 1967 when Apollo exploded on the launch pad itself, instantly incinerating 3 Astronauts, a 21 months' delay ensued. The same year the Soviet Cosmonaut, Komarov, plummeted to his death in Soyuz I. This stalled Soviet Banned flights by 18 months. Anyway, the forthcoming Jupiter missions will have to be shelved till July 1987, to ensure the proper alignment of that planet with the Earth and the Sun.

The catastrophe completely through off gear a crowded shuttle programme. First, following a ruthless investigation, the shuttle should be pronounced space worthy. Next, another shuttle would cost 2 billion dollars. Lastly, till Challenger is replaced, the commercial airline schedule like tight-rope programme, which encompasses several satellite launches for various agencies on a commercial basis and umpteen other experiments, will have to be executed by the 3 remaining shuttles. The future shuttle itinerary is indeed foggy. More importantly all this would retard America's grand plan of floating permanently manned stations by the mid- nineties. Such Space stations were to be assembled from modules ferried up by the shuttles.

Meanwhile preliminary investigation has thrown up a shocking fact: The shuttle tragedy was entirely avoidable. It appears that the blow up vas triggered off by the malfunction of one of the synthetic rubber gaskets that soul segments of the booster rockets: That the material of those so called 'O' rings is not resilient enough at low temperatures had rendered them suspects. Apparently one of these 'O' rings had sprung a deadly leak. Because, the temperature on that wintry launch day in Florida was indeed, alarmingly low. Lower than at any previous launch. It looks like the shuttle was blasted off after strong objections raised by technical personnel were overruled by NASA authorities.

Does this reflect the new ethos of Space Commerce? Interestingly, high quality microscopic latex spheres for the calibration of various instruments, were fabricated in conditions the near zero gravity conditions of outer Space, under the shuttle programme. They were commercially marketed last hour. The first outer Space products on sale down below!

There is a story that when reporters asked the would be first American Astronaut in space, John Glenn, how he felt, he quipped that he was very nervous, because, the space craft had thousands of parts, each supplied by the lowest bidder! That was the era of urgency, spawned by political prestige. However, the underlying principle had been safety. As Von Braun, a doyen of the America Space programme, had underscored, each of the important parts had to be duplicated, A double guard against malfunction. But probably, the days of Space program solely for science or political prestige are over.

7
PRECEPTORS OF THE GODS

An ancient Sanskrit couplet proclaims: "Those who know the time, the planets and the constellations are themselves Preceptors of the Gods?". A question still intriguing, still unresolved is, "who were the earliest Preceptors of the Gods?"

The need to compute the time of the day, month or year can be tracked down to the roots of civilization itself. Visualize a freezing earth, enormous tracts, groveling under ice, people migrating in herds, seeking shelter in caves and hunting when hungry, in ceaseless quest for more congenial climes.

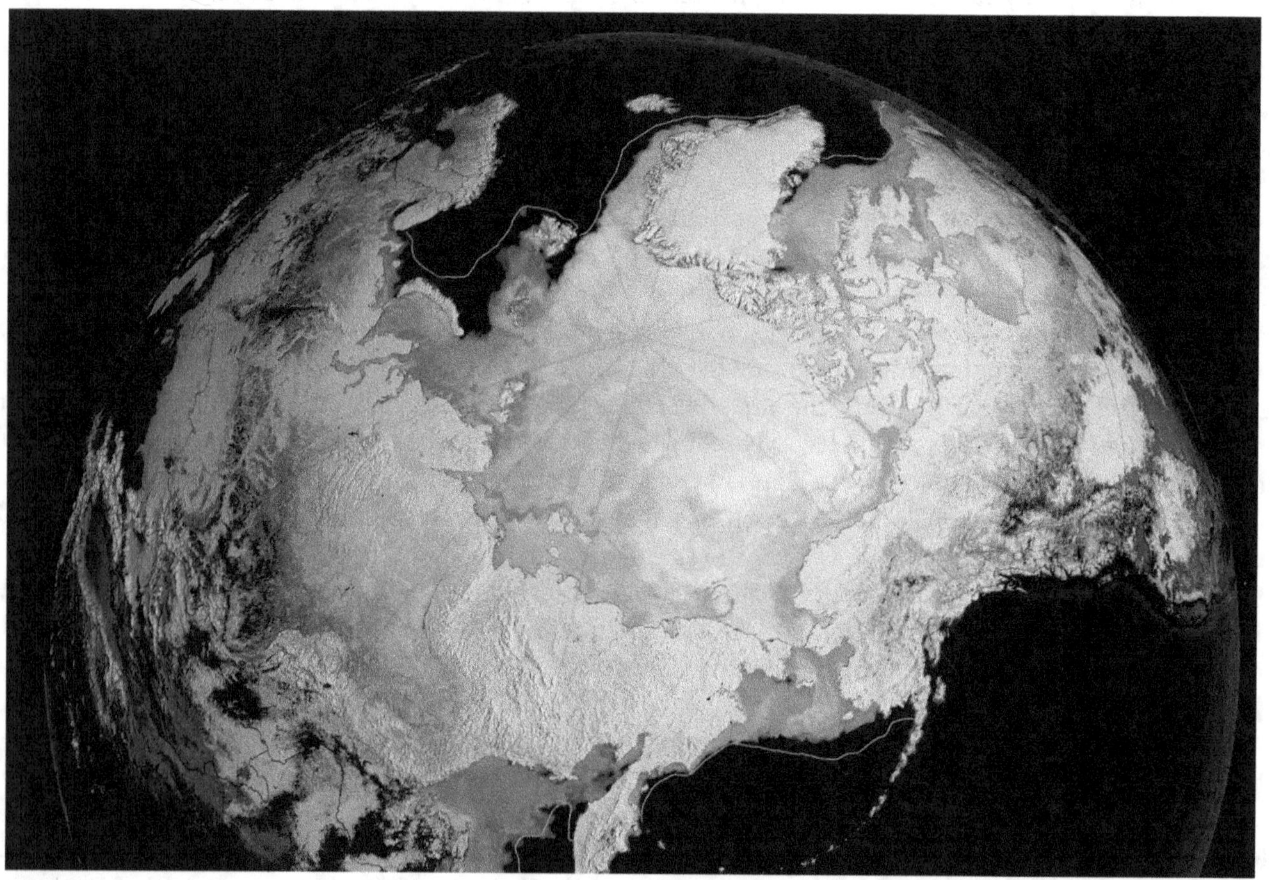

Earth covered in ice (NASA)

Then, the last of the past ice ages, thaws. The earth bares her fertile bosom. The tribes abandon their nomadic habits and impromptu huntings. Agriculture is born. This moment is snapped up on the Visnu Purana. The transition from a hunter gatherer society to an agrarian one with commerce and what not.

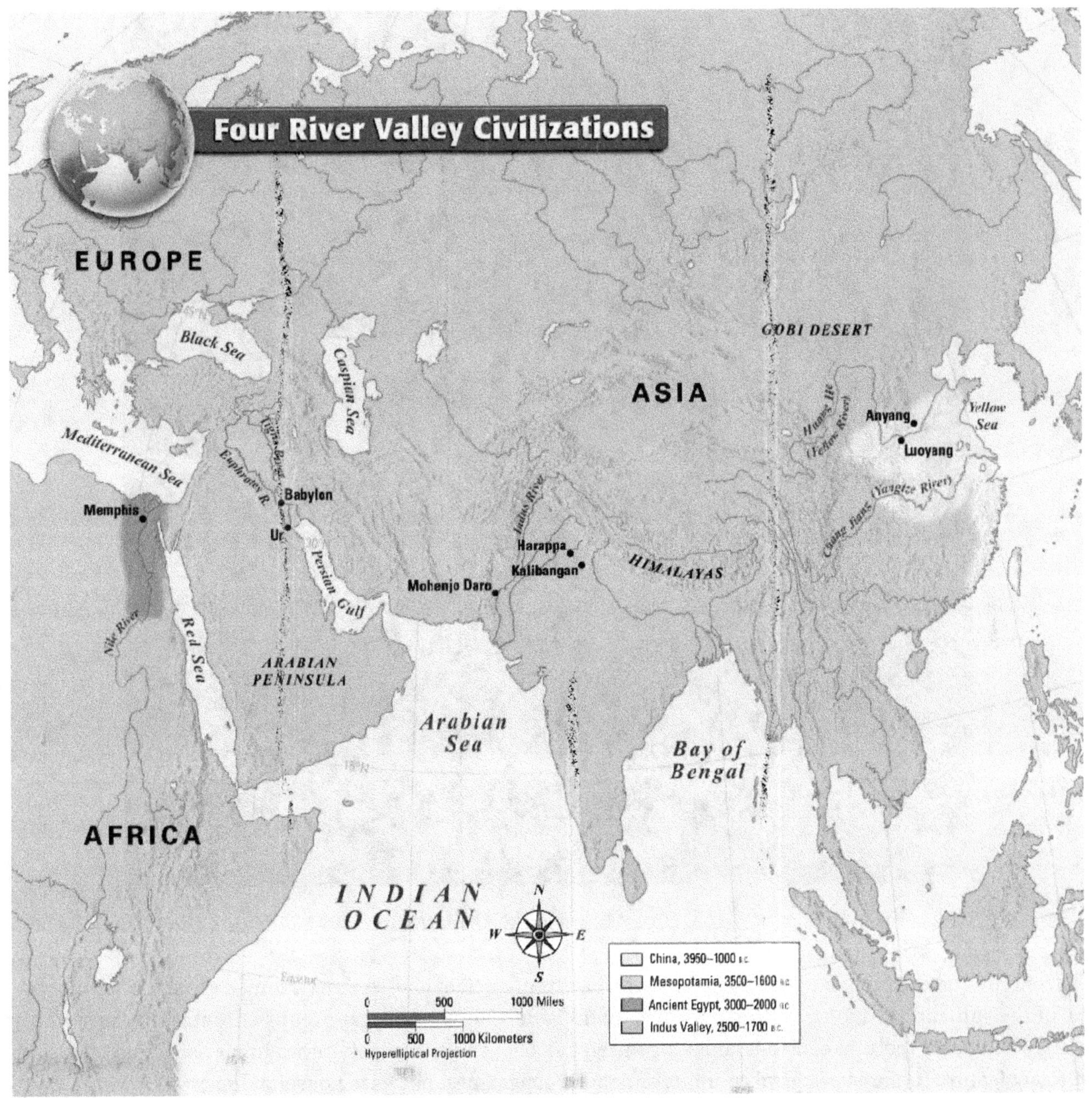

River Valley Civilization

That was maybe, 10,000 years ago. Soon, man was desperately seeking a clock-cum- calendar to regulate agricultural and growing cultural activities, and, succor slighted from the heavens. Literally: He discovered the clockwork dram in the high skies. The rising and setting of the sum and other celestial objects threw up the day. The span of 29 1/2 days separating successive full moons, yielded lunar months of alternately 29 and 30 days. The moon itself circles the earth in 27 1/3 days. Finally, the sun glides through the stars, apparently emitting the earth in about 565 days, a year about 11 days more than 12 lunar months The seasons repeat themselves after a year.

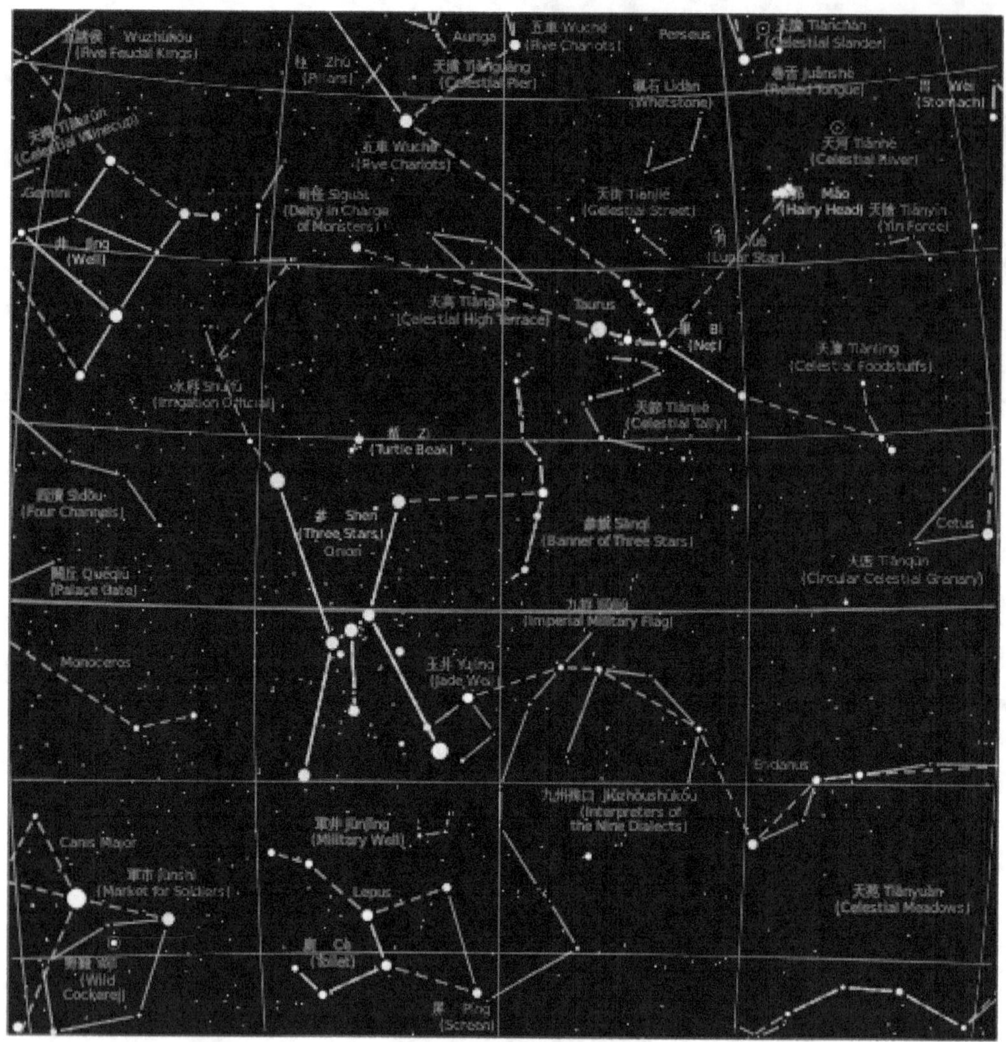

Lunar Asterisms (Nakshatras)

Astronomy and the calendar were born. The high priests of ancient civilizations scanned the sky. They plotted the paths of the sun and the moon, tracing out constellation and asterisms or star-groups enroute. This age-old legacy persists. The twelve zodiacal constellations or sun-signs (Rasis) and the twenty-seven lunar asterisms (nakshatras), even today feature in the astrological columns of popular magazines. They are household names. However, mystery obscures their exact origin.

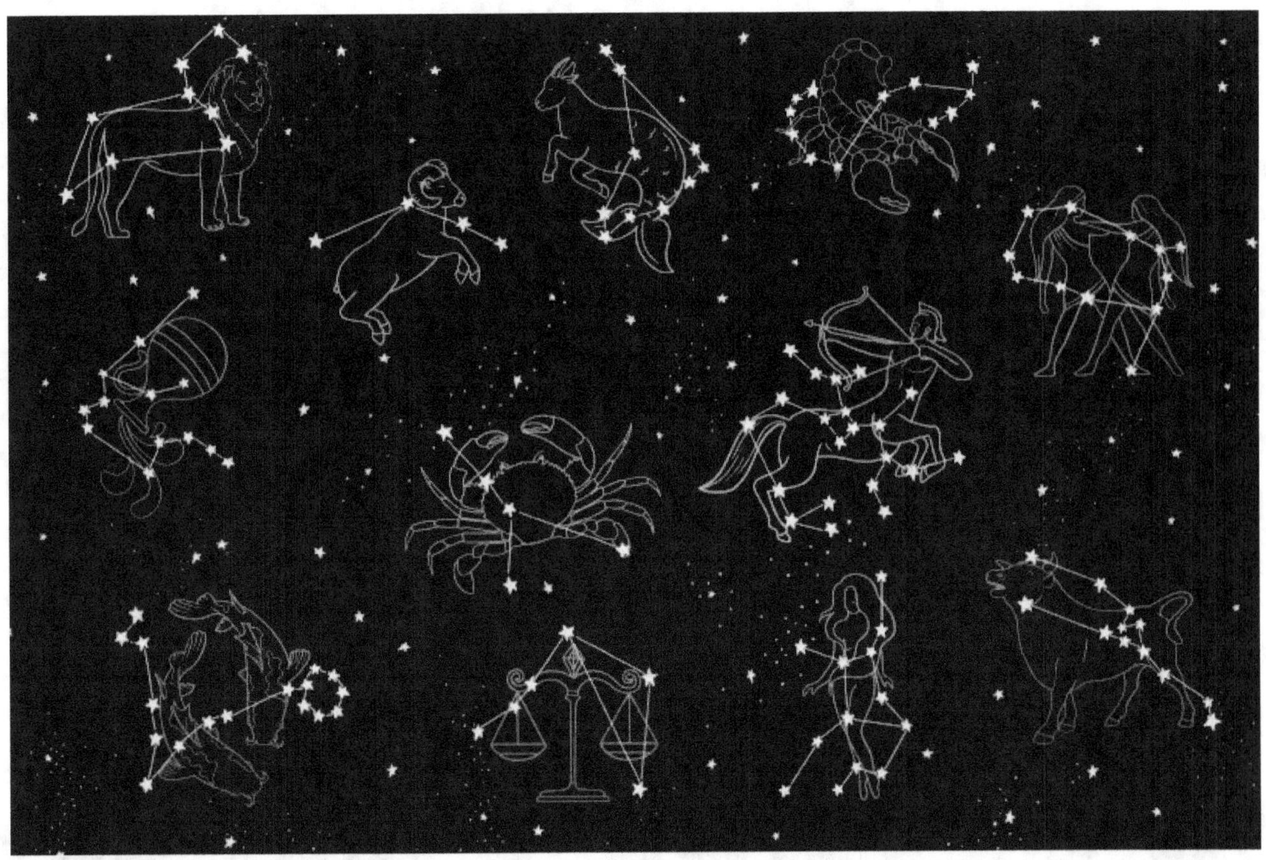

Signs of the Zodiac

Some 5000 years ago, the ancient Sumerians, flourishing between the Tigris and Euphrates rivers, employed a space, if workable, luni-solar calendar of obscure origin They enumerated 12 lunar months. To reconcile the gap with the sun's year, they threw an extra sooth every three or four years. The Egyptians on the hates of the about 4000 years back, ignored the moon and worked out a calendar of 12 thirty day months, appending the 5 odd "Leap days" at the end of each year.

By about 7000 B.C., the Babylonians enshrined centuries of Babylonians in an astronomical treatise, Mul Apin. This described the motion the on, the soon and sere the five naked eye planets. 13 or 14 close by constellations also but not yet the twelve sun-signs. They had or refined the Sumerian calendar to nicety.

Mul Apin (source the British museum)

The sun-signs first shown in recorded texts only around 400 B.C. By then the Greek thinkers were flourishing. Thales of Miletus, Pythagoras of Samos and others had venured afar, Into Babylonia, and probably India They imbibed and improved Asiatic astronomy and mathematics. Alexander too, battled his way right to Indie, transporting beak piles of Babylonian clay table and Asiatic concepts. This unfurled a new epoch. The Greeks ploughed back practically second-hand astronomy and mathematics into Asia. Gradually, a hybrid astronomy evolved in India. then the uncanny mathematicians, fashioned an end product that excelled the Greco-Babylonian system.

Historians conjecture that the 12 sun- sign evolved sometime during the Greco- Babylonian Interaction, around 400 B.C. But could this conclusion be too glib? A few hymns from the Rig Veda and later Vedas certainly hint at lurking mystery The sun-signs might have been invented much earlier.

The Rig Veda, a compendium of the oldest Hindu, even Indo-European hymns, can be traced back to, at the latest, 1300 B.C., conservative Western scholars favor this date. Thousands of the original hymns are conjectured to have been lost over the mile. For example, a Rig Vedic hymn in the Sama Veda incorporates a stanza extolling the Aswin twins, not found in the present collection.

In the Mahabharata, a discouse of the sage Dhaumya chants one such lost hymn, extolling the Aswins. This is very suggestive. "... The wheel of time represented by the year has a navel represented by the six seasons. The number of spokes attached to that nave is twelve, as represented by the twelve signs of the zodiac..."

A Rig Vedic hymn refers to the 12 fixed parts of the holy wheel and also to the 12 solar forms probably the origin of the later, Puranic concept of the 12 Aditya's or sun gods. These 12 divisions refer to the solar divisions and not lunar months. Because, the same hymn later describes the 12 lunar the and the trailing 13th, extra month. Prof. Wilson Invokes the 12 zodiacal signs to Interpret another hymn. Other Vedic texts also reiterate all in. for mole, the Atharva Veda. The deeper significance to that both the luni-solar and solar calendars were in vogue in the Vedic world.

If the Vedic Hindus did indeed fashion the zodiac, how could the Babylonians or Greeks contact these concepts? Well, plenty of cross-currents lashed the ancient world, too. Even trade with the Sumerians, calendar and clay tablet inscriptions unearthed in Turkey in 1907 bear testimony to this. These celebrate a truce between two warring Aryan tribes, the Hittite and the Mitanni, around 1400 B.C, at the Hittite capital of Boghozköy or Hatusha near Ankara.

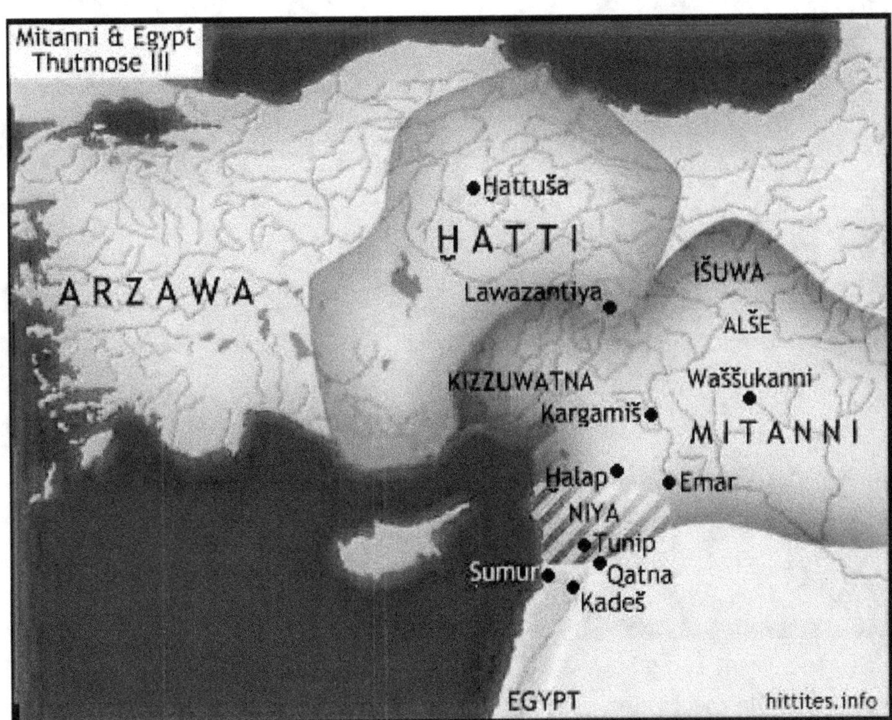

Map showing the locations of the Hittite Empire and Mitanni Civilizations at their height

1380 BCE, 3400 Years Old Inscription of Treaty Between Hittite King Suppiluliuma I and Mittanni King Shattiwaza Invoking Hindu Gods Mitra, Varuna, Indra and Ashvins as Divine Witness. This Inscription Was Found In Bogazköy, Turkey. British Museum

The Lion Gate at Hattusas in Boğazköy, Turkey.

Clay tablet with Sumerian cuneiform inscription ca. 2000B.C

It is startling that the deities invoked as witnesses are the Rig Vedic Varuna, Mitra, Indra and Nasatya (Asvins). Also, between 1746 B.C, and 1180 B.C., an Aryan tribe, the Kassite swamped Babylonia. The Kassite deities are again, astonishingly Rig Vedic. For example, their chief God, Shurias (Surya), or, Marytas (Marutas) and so on.

The History of the lunar asterisms or nakshatras, is fascinating, too. Probably, 28 had been originally calibrated along the moon's path. Later, this was corrected to 27. One of the later Rig Vedic hymns, refers to nakshatras: If Agha and Arjunis are identified with the nakshatras Magha and Uttara and Purva Phalgunis. Curiously enough, Uttara Phalguni was the birth asterism of the epic hero, Arjuna. "... so, they know me as Phalguna", Arjuna divulges in the Mahabharata. Further, the Rig Vedic Tishya could be identified with the Black Yajur Vedas Tishya, that is the nakshatra Pushya.

An ingenious theory links the three cosmic strides of Vishnu or the Sun, as extolled in the Rig Veda, with the 27 nakshatras. This concept, Incidentally, inspired the later Puranic legend of the King Bali, the Vishnu avatar, Vaman, and the three cosmic strides. If 12 sun-signs exactly measure out 27 nakshatras, then each sun-sign encloses 2 1/4 nakshatras. The three steps of Vishnu would correspond to the exactly 5 points of the zodiacs, where a whole number of nakshatras exactly coincide with a whole number of Rasis. This interpretation would establish the Rig Vedic origin of the sun signs and lunar asterisms.

This is mystifying. Because, the later Atharva Veda enumerates, probably the more antique 28 nakshatras, the extra nakshatra being Vega (Abhijit). But then, the White and Black Yajur Vedas recognize 27 nakshatras.

However, the oldest text with a purely astronomical and calendrical theme, the Jyotish Vedanga, spells out 27 nakshatras. This post Vedic treatise was composed about 1350 B.C., by Suci, probably the Sovereign of Magadha, under the tutelage of his preceptor, Lagadha. Its impact lingered till about the third century A.D. By that time Greek concepts were already diffusing into India. In the Mahabharata times, probably around the compilation of the Jyotish Vedanga around 1350 B.C., the hoary nakshatra system held sway.

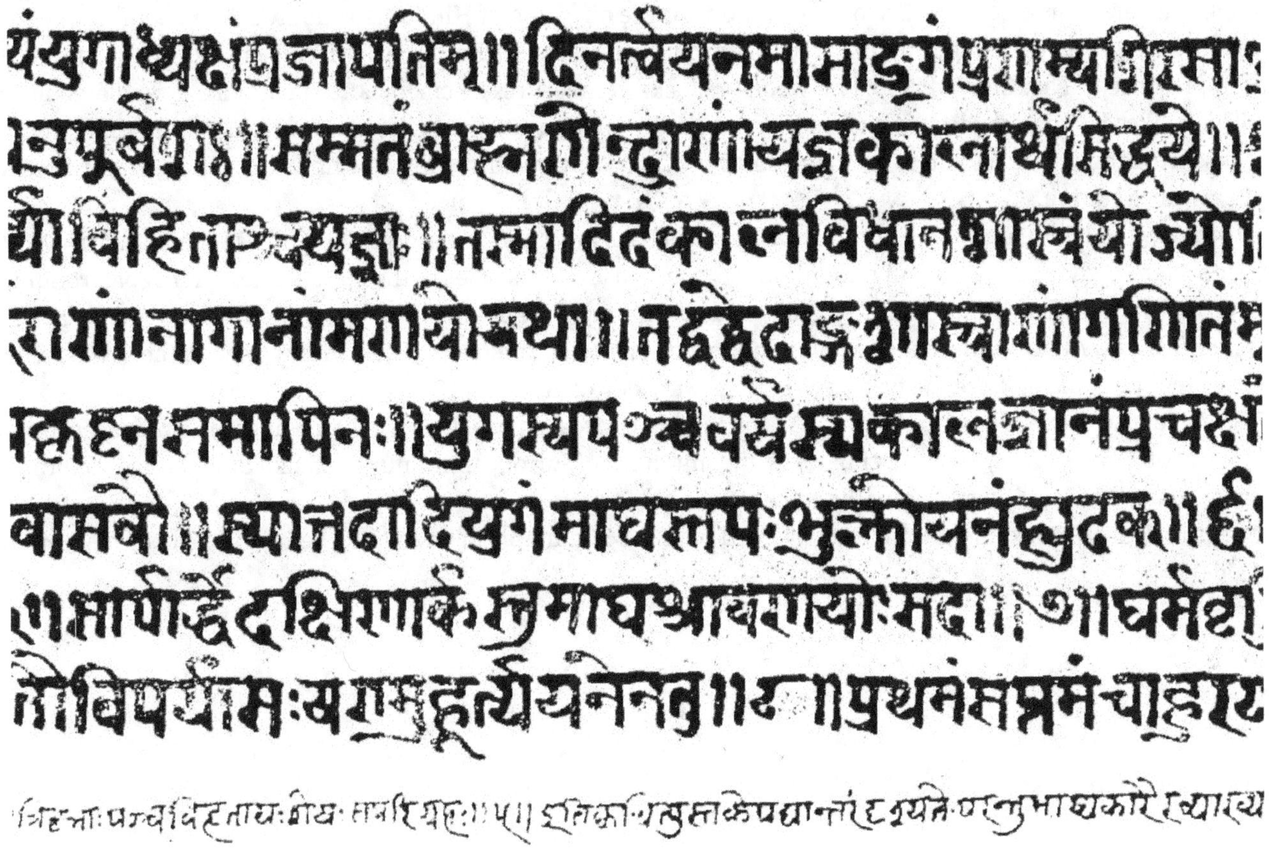

An excerpt from the Jyothish Vedanga

The tussle between the 27 and the 28 nakshatra lunar zodiac is triggered off by the fact that the moon orbits the earth in 27 1/3 days. Earlier astronomers erroneously computed the period as 28 days. They delineated as many nakshatras, the moon transiting one nakshatra per day. Later they slashed the extra nakshatra. An allegorical Puranic legend vividly portrays this conflict. In the Puranas, the nakshatras are daughters of Daksha, maybe the earliest astronomer who located them. From the usual tangle of various Puranas, a curious and coherent tale emerges: Daksha married off 27 of his daughters to the moon who sojourns for a night with each spouse. A 28th daughter, the superfluous nakshatra Sati, was wedded to Siva. It was a spectacular sacrifice, Daksha was compelled to disown the uninvited Sati. Disconsolate, Sati immolated herself while Veerabhadra, the fiery creation of the furious Siva, beheaded Daksha and installed a goat's head on him. Interestingly, the same mood is echoed in another, probably equally allegorical Chinese legend. That of Hi and Ho, two inexpert, imperial astronomers of the Emperor Chin K'ang. In 2136 B.C., they failed to predict a solar eclipse and were executed.

It is remarkable that the Chinese too, employed a 28 nakshatra zodiac, the Hsiu.

Earlier by about 1400 B.C, they plotted lunar asterisms, while the Tom Ling, around 850 B.C. spells out 23. This kicks up the question, the lensed the nakshatras?" Many favor Indians, some suggest Chinese, Babylonians too, here he proposed unconvincingly.

However, the Vedic hymn, the Jyotish Vedanga and the Daksha legend, all indicate an Indian origin. And, India-China cross-currents did exist, even prior to awed Chinese travelers like Fa hien, or Yuan Chuang. Or the envoys of the Emperor Lalitaditya of Kashmir, who were dispersed afar into China, in the sixth and seventh centuries.

In recent times, two curious and conclusive clues have surfaced:

Around 1890, a very antique manuscript was unearthed at Kugiar, near Yarkand in Sinkiang province. In the archaic Gupta it describes the 28 nakshatras of the Atharva Veda, as expounded, in an earlier epoch, by Pushkarasari. This hoary Indian scholar has been mentioned in old Hindu writings. There is even the possibly that he lived in the Mahabharata era. This discovery discloses a window for the diffusion of Hindu Astronomy into ancient China.

And then, recently, an old star map vas salvaged from a Pagoda at Suzhou, one of the ancient cultural centers in the Jiangsu province of East China. Printed on wooden blocks, it measures 25 cm. by 21.2 cm and has been dated at 1005 A.D. This makes it China's oldest star man. It antedates by about a 100 years the previous oldest star chart that had been discovered in 1971 in a tomb at Xuan ha, northern China. Interestingly, the new find depicts the 28 nakshatras in a Sanskrit Incantation.

Obviously, the Hindus invented the 28 nakshatra zodiacs and then revised it. The Chinese merely borrowed the archaic system without even critically examining it.In any case all this pales in the light of latest archaeoastronomical discoveries of the author. To sum up, The Vedas and astronomy were composed at least about 14000 years ago in the vicinity of Turkey. All this is descried in detail in the author's celebrated book, The Celestial Key to the Vedas, (Simon & Schuster), one could also see the website, Veda.bgs.webnode.com

The ancient scenario is obscure. As will be seen, history conceals more mystery than what scholars suspect or even like to accept. For instance, recent estimates of the antiquity of European monuments like Stonehenge, debunk the current theory that Mesopotamia was the epicenter of civilization.

And, the Vedic Hindus might well have invented both the Lunar and solar zodiacs, being the earliest Preceptors of the Gods.

8
THE BLINKING CRAB

In the early hours of the fifth of July, the Chinese peasant was wading through soggy paddy fields. He glanced at the eastern sky. And stood transfixed.

In Arizona, the other half of the world, an Anasazi Indian boy stumbled out of his wigwam. He gazed up and gasped. Then, spinning around, be sprinted towards the medicine man's hut, yelling incoherently.

A freak spectacle had spangled the sky. To appreciate this celestial drama, you must first visualize a colossal cloud of Hydrogen, a few million times as massive as the earth and a thousand million million million million times as voluminous! Hydrogen of course, is the simplest element. Just an electron orbiting a proton. The cloud is chillingly cold and blindingly black. Then, the mammoth begins to collapse under its own weight, warming up over millions of years. Its surface begins to glow gently while the core is ablaze.

Suddenly there's a miraculous transformation as the core temperature soars to a stupendous fifteen million degrees centigrade, unleashing thermonuclear fusion: Every four atoms of Hydrogen are rammed in and an atom of Helium emerges. In this process, the core disgorges more energy in just one second, than can be gobbled by energy-greedy mankind in a thousand years! This energy gushes out and halts the cloud's collapse.

A star is born: A huge ball of seething glittering Hydrogen, over a million times the earth size. It adorns the heavens for up to a few thousand million years. Meanwhile, all the Hydrogen fuel in the core is spent up. No more energy surges out. And the star collapses all over again. The core temperature now zooms to an inconceivable hundred million degrees centigrade. This triggers off a new nuclear reaction. Three atoms of Helium are fused into an atom of Carbon. The energy vented in the process stalls the collapses again. A series of such collapses yield Oxygen, Magnesium and more complex elements. Finally, when this seethes at an incredible thousand million degrees centigrade, Carbon is transmuted into iron.

Then there is a dramatic turnabout. Nuclear reactions involving iron severely absorb energy. This reinforces gravitational collapse. The core caves in catastrophically. The resulting shock wares violently blast off overlying layers of the stars.

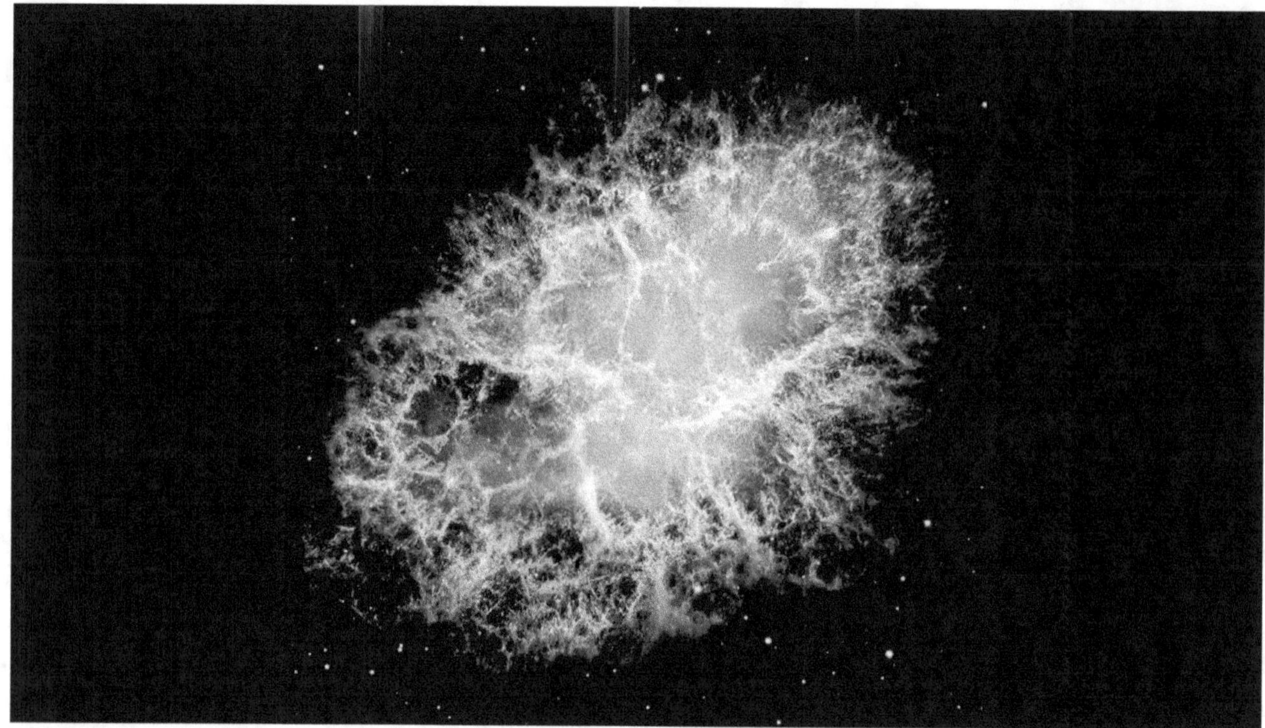

This Hubble image gives the most detailed view of the entire Crab Nebula ever. The Crab is among the most interesting and well studied objects in astronomy.

This image is the largest image ever taken with Hubble's WFPC2 camera. It was assembled from 24 individual exposures taken with the NASA/ESA Hubble Space Telescope and is the highest resolution image of the entire Crab Nebula ever made.Credit:NASA, ESA and Allison Loll/Jeff Hester (Arizona State University). Acknowledgement: Davide De Martin (ESA/Hubble)

To visualize this cosmic explosion, first picture the blow out of thousand of dreaded Hydrogen bombs. Enough to wipe out humanity. Such a stellar star outbursts would out do a million million million million Hydrogen bombs erupting in unison! If one of our closer stars blew out . the earth would be wiped out!

Some scientists propose that that's exactly what happened about sixty million years back. A few hundred million milion kilometers away. The floods of instant cosmic rays spewed out, ravished the earth's atmosphere and distorted the vital ozone layer which shields us from the sun's lethal ultraviolet radiation blast. This finally exterminated, in one swipe, the whole Dinosaur population. Isn't life perched on a cosmic razor's edge?

In the universe, such stellar blow outs are a commonplace. Millions per year. But -thank your stars -most stars are so distant that these mind-blowing events go unnoticed. Or, they surely deck the sky with a dazzling new star, a supernova. One of the last super nova witnessed was in 1604, by the renowned astronomer Johannes Kepler.

That is what arrested the attention of the Chinese around the fifth of July, 124, as records of the Shing dynasty reveal. They dubbed it a guest star. It adorned the constellation of Taurus, or Rishabh, out-dazzling for three weeks, all but the brightest. Probably the brightest object ever glimpsed in the sky. Some allege that you could have even read by its light at night. Flashing near the rim of the crescent moon, it captured too, the fancy of the Anasazi Indians of Southwest America. They preserved this stellar spectacle in rock paintings and carvings.

This Anasazi painting at Chaco Canyon may be a depiction of the 1054 supernova, which we see today as a supernova remnant, the Crab Nebula.Source:1054 Supernova PetrographImage Source
(http://www.astronomy.pomona.edu/archeo/outside/chaco/nebula.html)

Actually the star detonated about 4000 B.C., long before the ancient Egyptians had erected their gorgeous pyramids or the Sumerian astronomers examined the sky. It was so remote that its light-flashing at the ultimate speed of the universe, an unthinkable thousand million kilometers per hour-reached the earth after 5000 very years. Surely, such a cosmic catastrophe cannot peter out in a month long display of stellar sparklers. Could some trace persist?

Around August 1755, a French draughtsman cum part-time comet hunter, Charles Kessler, took a survey of the sky for an exclusive comet. He glimpsed it and then lost track. As he doggedly peered through his telescope, he thought he re-captured it in Taurus. Repeatedly. But unlike comets, this fuzzy patch didn't budge. Exasperated, anther colleague Messier compiled a catalogue of such nebulas or cloudy objects, to weed them out of future observations. He installed the nebula in Taurus at the head of the list of undesirables. Messier discovered an extraordinary number of these objects. Ironically, it was the Messier Catalogue of nebulae that earned his immortality.

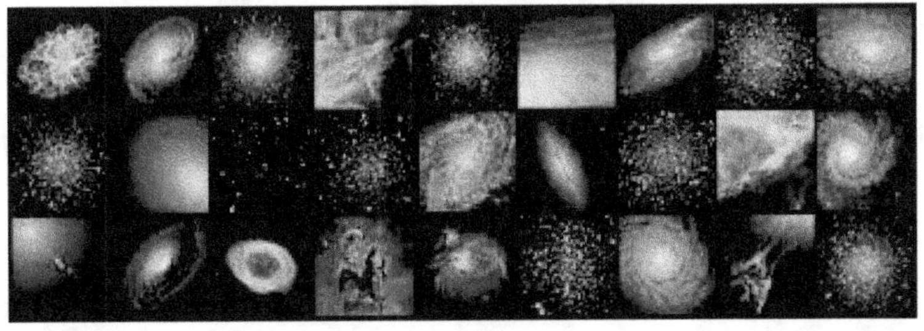

The Messier catalogue (NASA)

Then, in 1844, the Irish Earl of Hoss, scanned the sky through his powerful telescopes scrutinizing, not comets this time, but nebulas. Particularly Messier 1 alias the nebula in Taurus that annoyed Charles Kessler. He sketched this extra-ordinary spectacle, with tentacle like filaments fanning out. Rather like one of these filamentary clouds you can see so often, it resembled a Crab. It was rechristened, "The Crab Nebula".

Photographs over even twenty year spans reveal that the cloud is still swelling furiously. The gases are hurtling outwards at average speeds like a 1000 kilometers per second. So, working backwards, astronomers computed that the sprawling Crab nebula would have appeared as a compact star a little over 900 years back. They then delved into antique records and discovered the guest star that had astonished the Chinese, of course! At the same spot too. Undoubtedly, the Crab Nebula is the debris of the stellar blowout witnessed in 1054.

The Crab Nebula has continued to fascinate and tantalize astronomers. It is a treasury of cosmic surprises. First, photographs threw up a faint central star, closely guarded by the crab Nebula. With a staggering surface temperature too, of half a million degrees centigrade. An enigmatic almost impossible object. This star fires the Crab Nebula, illuminating it. By the mid-sixties, astronomers were speculating that it was a "white dwarf"; An intensely hot star, with a sun like mass, but squeezed into the earth's volume, a teaspoon of its material would outweigh todays giant aircrafts. Next, astronomers discovered that the Crab Nebula was one of the most powerful sources of radio waves.

Then, in 1967, astronomers struck a sensational discovery: Earth directed cosmic pulses or signals. Not the usual smooth stretches of radio waves. More like someone out there was flashing a Morse code with radio waves. Any one contacting us? Astronomers were breathless. They even tentatively christened the signal source, LGM 1: Little Green Men 1! Who says astronomers don't have funny bones?

Soon they detected such signals from diverse directions, surely, that many exotic civilizations couldn't be contacting us at once? Unless, so many space cosmic lighthouses, erected for the navigation of an ultra-spacefaring society were transmitting these flashes Reluctantly, astronomers banished these fancies and fantasies... They dubbed the invisible objects beaming these radio pulses, precisely 'pulsars'. As if you were on a beach and were getting the flashes from a rotating light house beacon!

Immediately, the Crab Nebula disclosed another of Its wondrous secrets. It central star was also flashing pulses. Light pulses. Thirty-three blinks per second. It was the first optical pulsar.

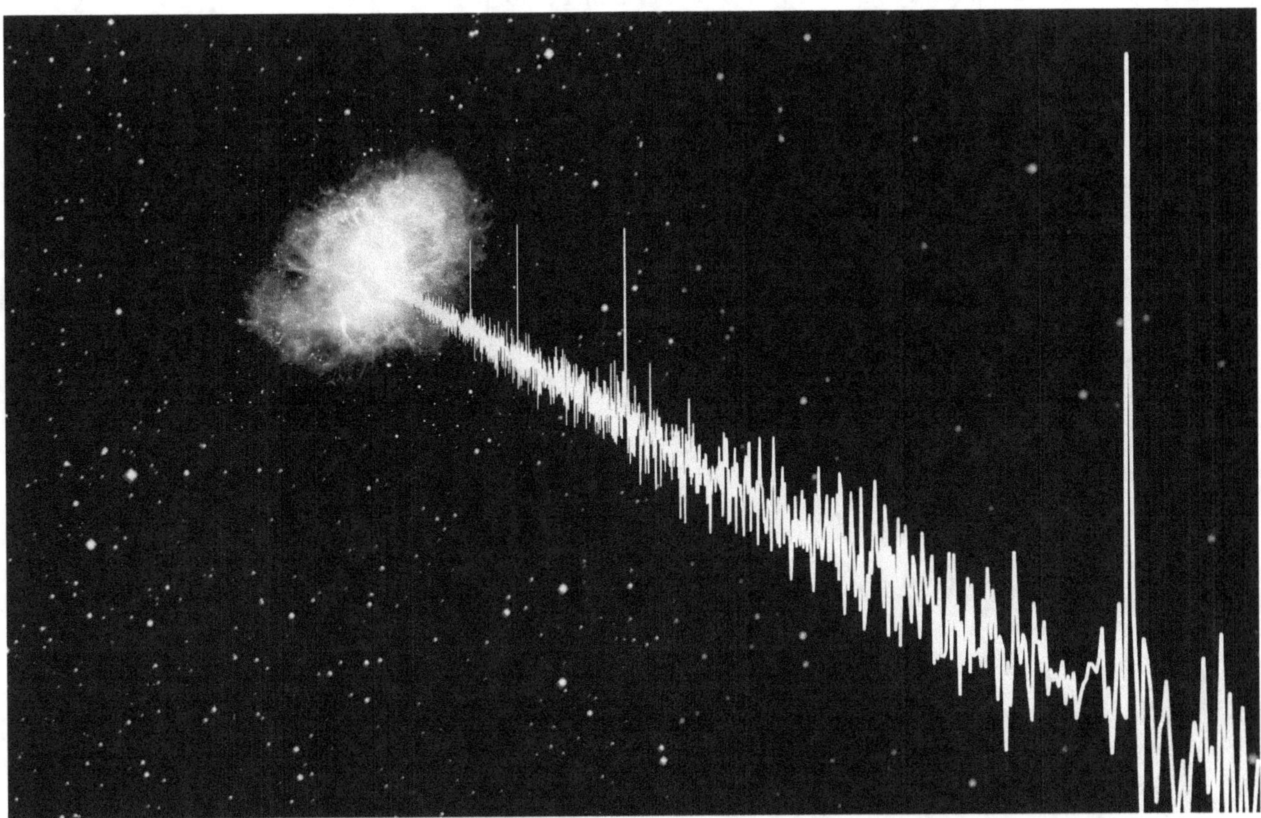

The Crab Pulsar and the Crab Nebula are one impressive pair, especially in this unique image made by a duo of NRAO radio telescopes.

These light and radio signals bring an exotic message. Not of remote civilizations, but equally mindboggling. Pick up the shreds and imagine a star that has erupted as a supernova, spewing out colossal amounts of Hydrogen, Carbon, silicon and what have you. The equivalent of hundreds of thousands of earths. If the remnant core is still over 500,000 times as massive as the earth, it would crumble gravitationally once more, to within an incredibly tiny radius --just about twenty kilometers! Due to its super compression, the electrons orbiting the protons of individual atoms would bash into the protons themselves, churning out neutrons: Matter as known would be shattered. The star would be one clump of super squeezed neutrons. Try conceiving of such a neutron star.... Just a pinch of salt from there as mentioned could only be hauled by a convoy of several diesel locomotives! And don't take that with a pinch of salt!

Neutron stars are swaddled by powerful magnetic fields, Electrons trapped in this magnetic web spiral about with awesome speeds, ejecting intense X rays, radio waves light and what not: Usually in a definite direction, rather like a spotlight. Moreover, the neutron stars' spinning at dizzy rates. The effect is that of rotating light house beacon. The radiation spurted out would strike the earth once per rotation, that is, is pulses, As the beam flashes past: Pulsars.

However, pulsars play a losing game. They leak energy by rotation and radiation. Their spin is slowing down. The pulses are progressively delayed at predictable rates.

The Crab pulsar, for example, is slowing down by one millionth of a second per month. It is the fastest spinning pulsar to date. And the youngest too. The Apollo astronauts detected X-rays streaming from it. It vents gamma rays and infrared or heat waves as well. Astronomers speculate that it could also be a source of the hypothetical gravity waves. Observations are under way to pick them up. According to Einstein's General Theory of Relativity, gravity waves are released by the motion or collapse of massive bodies. So the Crab pulsar could well be another touchstone for Einstein's ideas. In a recent presentation, the author depicted a scenario where an astronaut was blown off by an asteroid hit and was utterly lost in space Accidentlay he could latch on to a pulsar light house and then he slowly works his way into the inner solar system! Saved by the blinking crab!

Pulsars have flashed us the incredible saga of inconceivable events that rend the deep interior of stars. Now they are transporting an equally astounding message to the stars. A tale of life, intelligence and civilization. The space crafts Pioneers 10 and 11, the first manmade objects to hurtle out of the solar system, to the stars and beyond, are conveying a plaque each. These depict an array of fourteen pulsars with their rotation rates. Millions of years hence, this pulsar map, sort of a stellar signpost, would enable aliens to track down the sun. And, from the spin rates, they could compute the epoch of the earth's ephemeral existence.

9

EPOCH OF THE EXTRA-CELESTIALS

The terror-struck moon man gaped at the black sky. The glinting dot swelled and swooped. Jerkily it landed. The magic ship of the evil star Gods, displaying its starry banner. The Evil One emerged, with his insignia of stars. The moon man trembled as the God approached springily.

Suddenly another spark glittered in the black background. A dazzling craft alighted. From the good gods of the crescent earth. The Benevolent One clambered out, flaunting his insignia --- the crescent and the awesome hammer. Both Gods confronted each other, wielding terrible weapons. There were sudden blinding flashes. Then both vanished.

The moon man shouted with joy. "The Oracle's prophecy has been fulfilled of the crescent earth has been rescued from the moon from the designs of the Evil God of the stars!"

This could be an excerpt from a future comic strip looking askance at the primitive phase of the space age. Consider. The world witnessed an almost equally ominous and symbolic space chase. Remember, the Apollo 11 astronauts hurtled behind the speeding, mysterious Soviet Luna 15 space-craft, with its top secret mission? And, quite accidentally, as the world paused breathless, Luna 15 crashed onto the moon, the very day Armstrong and Aldrin alighted there.

This was the crescendo of man's most magnificently orchestrated project yet: The then thirty billion dollar programme encompassed 20, 000 industrial plants and 200 American academic institutions. It harnessed an over 40,000 workforce and millions of man hours, throwing up some 3000 wondrous technological innovations: From computers through new alloys and lubricants to super electronic systems and non-flammable fabrics. Even Teflon frying pans.

The trigger for this epoch launching project were two dramatic events. On the 12th April, 1961, the Soviet cosmonaut Yuri Gagarin was shot into an eath orbit. The first man to gate-crash into space.

Yuri Gagarin (ESA)

The then Soviet Premiere, Nikita Khrushchev taunted, "Let the capitalist countries try to catch up!" Within weeks, An American Atlas rocket, boosting an unmanned Mercury Capsule, veered off course devastatingly.

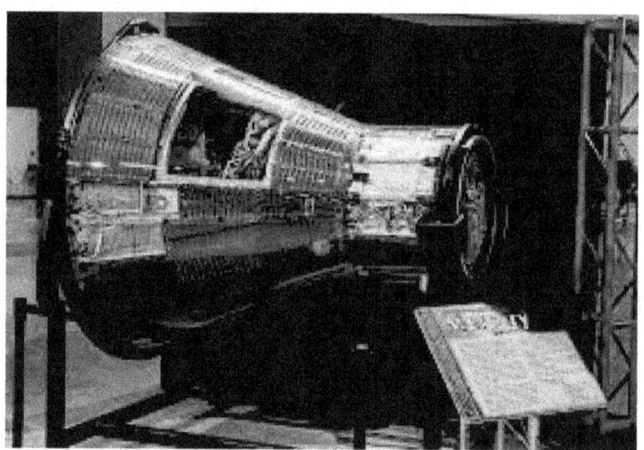

The Mercury Capsule (National museum of the United States Air Force)

The U.S. President, John Kennedy, propped up a down cast nation on the 25th May, 1961: "I believe that this nation should commit itself to achieving the goal, before this decade is out, of landing a man on the moon and returning him safely to the earth. No single project in this period will be more impressive to mankind or more important for the long range exploration of space and none will be so difficult or expensive to accomplish."

The decade of the moon had begun. By 1962, the blueprint for the assault on the moon had been delineated: LOR, the Lunar Orbiter Rendezvous technique. Three astronauts would be catapulted into a moon orbit. Two would breakaway in a smiler craft, the Lunar Excursion Module, swoop on to the moon, explore it and rejoin their orbiting companion. Then all three would be shot back to the earth for a splash down in the Pacific Ocean. Almost the entire U.S. space programme in the 1960s was governed by this overriding obsession: Man on the Moon.

The Mercury Project, in which single astronauts looped the earth, was trailed by the two man Gemini programme. The Gemini astronauts rehearsed diverse docking or link up manvoures, vital for the moon missies.

Gemini X command pilot John Young, left, and pilot, Mikel Collins, at Cape Kennedy (now Cape Canaveral) Air Force Station in Florida.Credits: NASA

On the side, the unmanned Surveyor and Lunar Orbiter probes relayed views of landing site candidates, and even examined them. In all these feats, the Soviets beat the Americans. But by progressively narrowing margins.

55 years ago Surveyor-1 makes a soft landing on the Moon (NASA)

Then, in January 1967, tragedy struck. Ironically, during a preliminary ground rehearsal of the Apollo moon ship. A fire blitzed through the space craft, incinerating within seconds, three astronauts, V. Grissom, R. Chaffee and B. While. Inevitable modifications ensued. All this stalled the Apollo programme for a year and a half.

The first manned Apollo flight came off only on the 11th October, 1968: Three Apollo astronauts orbited the earth for eleven days. Project Apollo was now neck to neck with time.

The three men who eventually formed the prime crew of Apollo 7. Left to right, are Walt Cunningham, Donn Eisele and Wally Schirra. Photo Credit: NASA

Meanwhile NASA was stuck with the vexing Lunar Excursion Module. This would ultimately ferry two astronauts to the moon. Worse, its crashes bugged the simulated tests. Neil Armstrong himself had a close brush with disaster, just about parachuting to safety.

To side step the adverse publicity churned up by these delays, the Americans shot the originally unscheduled Apollo 8, in December 1968. Three astronauts swung round the moon and triumphantly returned. Man's first voyage beyond the Earth and his closest look at a celestial world yet.

With a decades' end racing in less than a year away- NASA activity climaxed. Only two rehearsals for the Lunar Module had be hurriedly squeezed in: On the 3rd March, 1969, Apollo 9 executed the Lunar Module-Apollo craft docking maneuvers in earth orbit.

In May, Apollo 10 improved on the Apollo 8 feat, Stafford and German swooped to within 10,000 metres of the moon, In the Lunar Excursion Module before scooting back to the orbiting Apollo 10.

The stage had been set: From the billion dollar J.F. Kennedy Space Centre at Cape Canaveral to the 8.5-million-dollar Lunar Receiving Laboratory to which the earth coming astronauts would be directly transshipped. Just in case they transported back deadly lunar microbes which could hold entire humanity to ransom.

Finally, on the 16th July 1969, Neil Armstrong, Buzz Aldrin and Michael Collins clambered into the Command and Service Modules of the 375 million dollar Apollo 11 moon craft.

Neil Amstrong, Buzz Aldrin and Michael Collins

The world's most awesome rocket, the towering thirty storey Saturn V, blasted then off into history's most momentous odyssey.

The towering thirty storey Saturn V Rocket (NASA)

The first two stages and part of the third stage of the Saturn V rocket blazed, hurling Apollo 11 into a 188-kilometer-high earth orbit.

The Apollo 11 Blast off

Two hours and thirty-five minutes later, the third stage erupted again. The space ship shot to a speed of 38,691 kilometers per hour, bursting out of the earth orbit and plunging earh wards. It effected the third and final stage blasted them into a trajectory round the moon. Seventy-five hours after launch, hurtling across 380,000 kilometers, Apollo 11 looped into a 112-kilometer-high moon orbit. It whirled round the moon for about twenty-four hours, before Armstrong and Aldrin entered into the Lunar Excursion Module, Eagle, which delinked from the command and Service Modules.

Firing it's on rockets, Eagle true to its name swooped on to the moon in one hour and seventeen minutes. An incredible climax. Except that Eagle was plummeting headlong into a boulder ridden crater due to computer overloading! Proven cool head, Armstrong grabbed manual control and propelled the craft to the safety of the Sea of Tranquility. On the 21st July, after being cooped up in Eagle for about six hours, Arm strong shambled down the ladder. The mission was so slenderly optimized that the ladder would collapse in the higher gravity of the earth. Amidst soaring excitement, as he stepped on to the moon in his 100,000-dollar space-suit, well, Armstrong missed his lines. Instead of words carefully fashioned on the earth for history and progeny, he blurted out the equally historic, "One small step…"

Aldrin erects solar wind experiment (NASA)

After exploring the moon for two and half hours and twenty-one kilograms of moon rooks, Armstrong and Aldrin clambered back into the Eagle. Subsequently they soared up and docked temporarily with the Apollo craft, then the two moon men climbed back into the Command Module, they jettisoned Eagle, The Service Module's engines roared for the final act, slinging the Apollo craft earth wards. After a nearly sixty hour cruise the moon ship shed the Service Module and plunged to the earth in a flurry of parachute umbrellas. It splashed in the Pacific Ocean, Right on target. Amidst the dizzy euphoria of the fairy-tale finale, the crane commissioned to haul out the Command Module was suddenly adjudged unsafe. And the ten-million-dollar quarantine protocol was chucked overboard: The Command Module hatch was flung open in the Pacific Ocean Itself!

After Apollo 11, six other Apollo crafts were blasted off, as against the scheduled nine. Of these Apollo crafts made it back after slowly swinging round the moon under dramatic and gripping circumstances. Remember? in oxygen tank in the Service Module exploded due to overheating. Some of the later astronauts even managed to bump across the moon in reverse, and conducted diverse experiments. In all, the Apollo spacecraftsment transported back some 380 kilograms of lunar rock samples. The programme was abruptly cancelled with Apollo 17 due to a drastic cutback in funds. Just as the first legitimate scientist, geologist Harrison Schmitt, stomped on the moon, too.

Thus terminated an out of this world project. Launched for the wrong reasons and terminated likewise The handsomest feather in technologies cap, yet. Small bands of enthusiasts in the U.B. celebrate Space Week every July to fete the moon phenomenon. But ironically, intellectuals have denounced the Apollo programme no end. To be sure, the project has confirmed that the moon is lifeless: With the Apollo 14 mission itself, scientists realized that the exorbitant and elaborate quarantine ritual had been fashioned for non-existent moon microbes. Then, the moon harbors minerals of technological interest And it is different enough from the earth to debunk a common origin. It's distinct too from antique meteorites, the virgin stuff of the solar system. In fact, the moon, like the earth is a processed planet: The product of geological processes. An important hint this leaks out is that heavier planets may all have earth like interiors.

In anycase the Apollo eerily follows the Jules Verne's fictional scenario depicted in his nineteenth century novel But critics rail, couldn't this have been discovered at about one hundredth the cost with unmanned crafts?

a la the Soviet Lunas which returned moon samples? They aver that the Apollo project was a technological tour de force to prove America's pre-eminence in space against the backdrop of mounting Soviet successes and flagging

American morale. Even a political stunt to deflect public opinion from the ignominy of the Bay of Pig's misadventure, Prof. Carl Sagan maintains that the prime objective of the Apollo mission was never scientific. Even the landing sites were picked for astronaut safety. And then cynics sneak in with a "was a moon programme necessary to throw up spin offs like frying pans?"

Many critics berate the administration for tragically abandoning the moon project for the wrong reasons. The escalating economic burden of Vietnam for example. They even condemn the fast evaporating public interest and support, once America's paranoia had been bolstered. For example, TV stations in the U.S. intercepted sports programme to telecast live, moon walks. They were promptly deluged with indignant complaints. These critics also moan that the show room Saturn V s and Lunar Modules roaring and rearing to catapult the cancelled Apollo missions, are languishing as in museum pieces and tourist attractions in artificial moon dust.

The sudden and sharp fund and support cut backs have stumped hardy optimists like Carl Sagan and even safe-bet space prophets like Arthur Clarke confidently predicted permanent noon bases of the 1980s? Not even on NASA. But they might well be right! The current Artemis project is a back to the Moon Mission. That is another story though. .

10

WHEN THE SUN CHANGES ITS SPOTS

A leopard may not change Its spots. But the sun does, and though it is about 150 million kilometers away, that may affect you profoundly, Possibly in the net twenty years. No wonder, crores of ripens are being invested in solar research world over. In 1980, the U.S. even launched an exclusive solar space-craft, Solar Max. Some scientists are already a bit agitated.

Just consider, a twenty-two-year cycle is associated with droughts in the western U.S., as surprisingly, with sunspots.

Then, in 1979-80, when the number of sunspots was swelling to a peak, the eastern US enjoyed a benign winter, in sharp contrast to a couple of years earlier when the sunspot activity was lower. Suggestive?

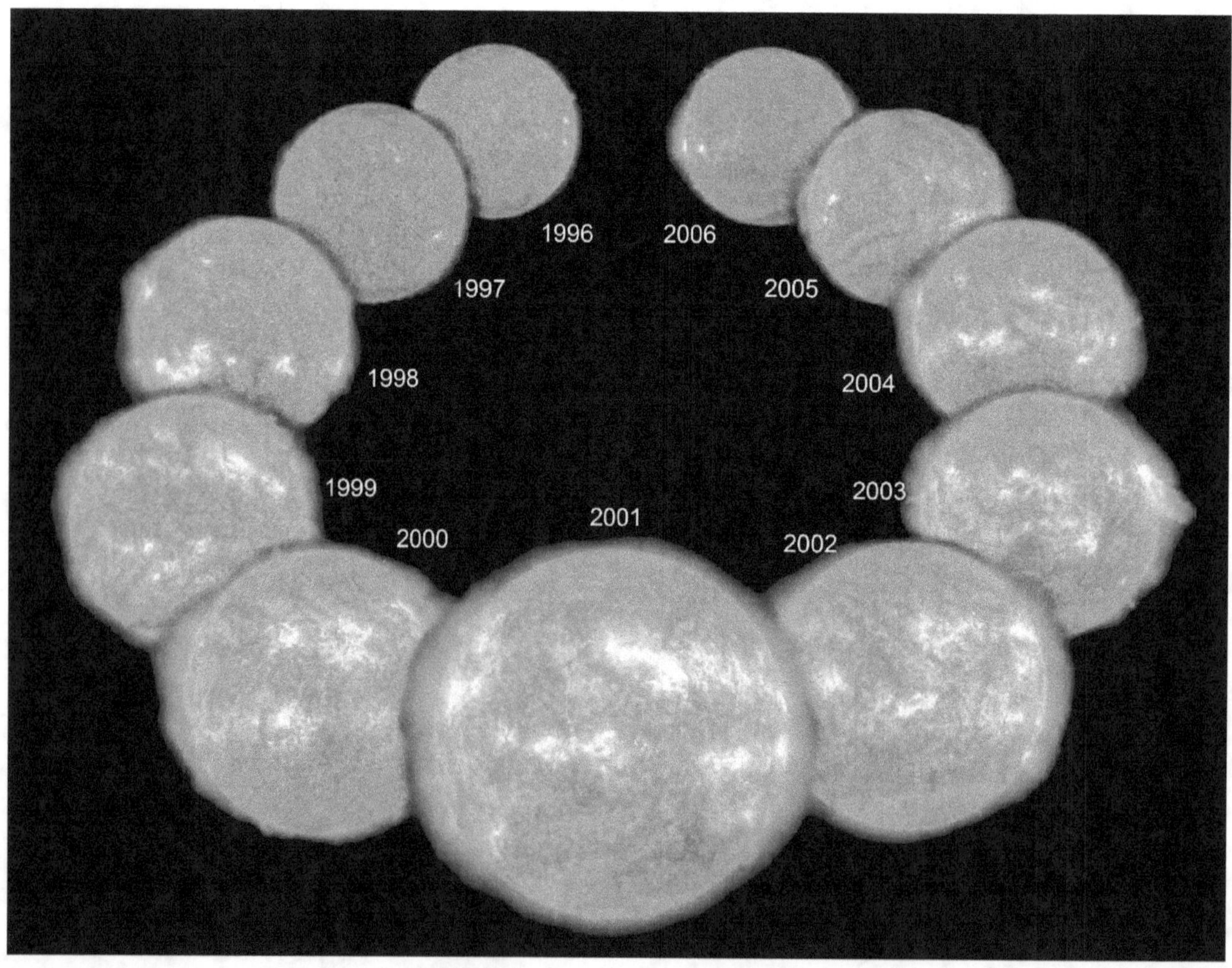

Increase and decrease of the Sunspots over 22 years (NASA)

People have detected sunspots for thousands of years-- as early as 800 B.C., according to one Chinese claim. But these dark dots and patches on the sun were mistaken for intervening objects like clouds. Could the perfect sun ever be scarred? You too can glimpse sunspots sometimes, at sunset or sunrise, when the sun is a genial pink. Around 1610, Galileo correctly concluded that they were part of the sun and the they drifted as the sun spun on its axis.

The apparently tiny sunspots could be a huge as the earth or even thousands of times larger. Though they look dark against the much brighter backdrop of the sun, they emit light. This light can be analyzed spectroscopically: It turns out that sunspots are linked with powerful magnetic fields and high temperatures like 4000 degrees centigrade. Modern theories associate sunspots with gigantic magnetic storms that are triggered off deep inside the sun and gradually surface.

Regions near sunspots suddenly brighten up so conspicuously that sometimes they can be seen against the sun's disk. These flares could be due to the explosive annihilation of the sunspots' magnetic fields. They could last for about half an hour, spewing out upto a thousandth of the total solar energy output, from a negligible fraction of the sun's area. During this time, starting with X-rays, a colossal flood of protons and other charged particles is disgorged. Part of this "solar wind" collides with the earth's upper atmosphere with diverse, weird results: radio-blackouts, telephone and, in general, telecommunication break-downs, disruption of satellite signals, the splendid aurora displays in the higher latitudes, ghost images on radar screens and even power cuts. During the second world war in fact, a deluge of such microwaves blacked out Britain's entire radar tracking network!

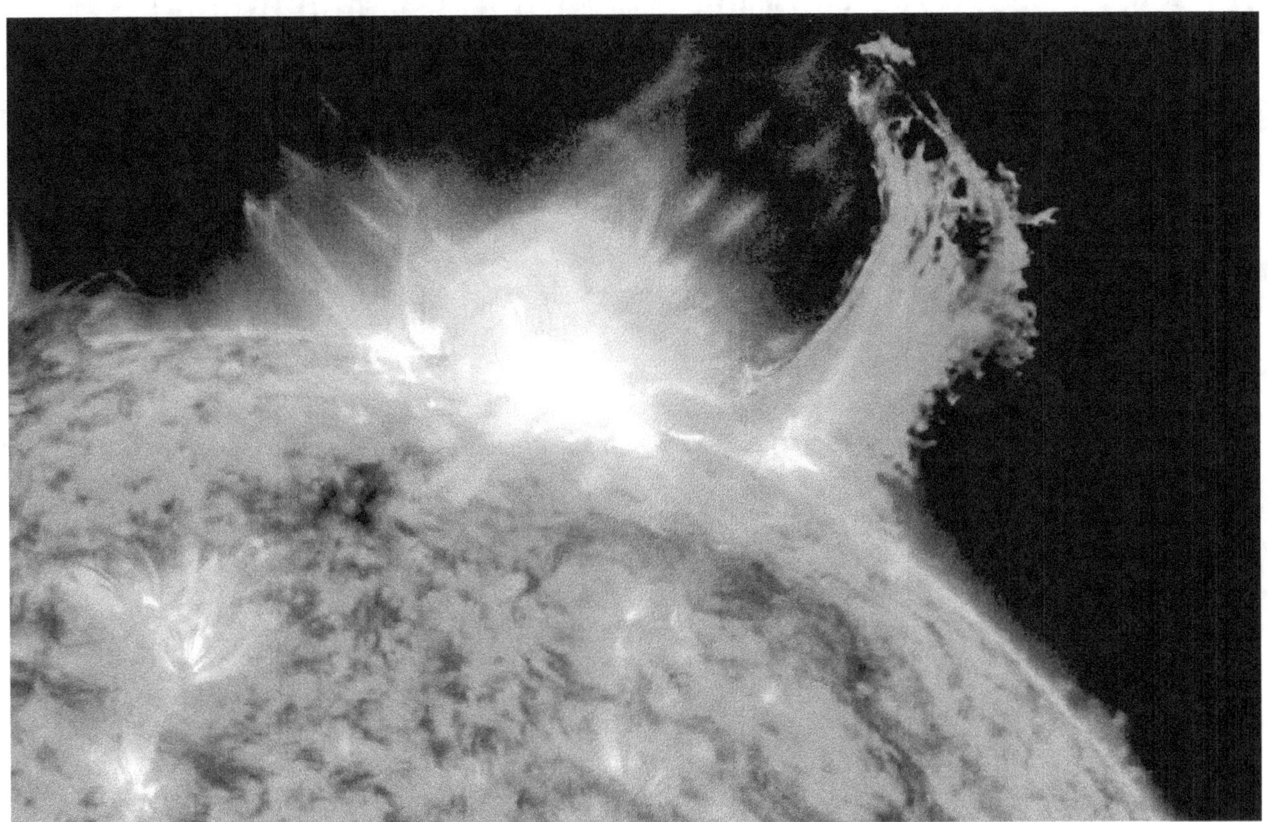

This image of a solar flare was captured October 2, 2014, by NASA's Solar Dynamics Observatory.

There are many, as yet inexplicable oddities about sunspots which may intrigue you. They tend to appear in pairs. One spot of the pair-- the one to your right, for example-- behaves like the north pole of a magnet, while the one to the left behaves like a south pole. Moreover, the right-left sequence of these north-south pairs of magnetic poles in the sun's Northern Hemisphere is reversed in the sun's southern hemisphere. The sunspots first appear about midway between the equator and poles of each half of the sun. Then they slowly spiral down and vanish near the equator after about eleven years. Again they surface, midway in each hemisphere of the sun, Only, this time there's a switch. The previous right-left sequence of north-south magnetic poles of the sunspot pairs on the sun's northern hemisphere, is transferred to the southern hemisphere and vice versa!

The eleven-year sunspot cycle was unexpectedly discovered in 1843 by a-- can you beat it ---German pharmacist, H.S. Schwabe. Of course, from 1826 he had been hunting for a non-existent planet that would crawl across the sun, showing up as a dot. So he had been religiously sketching the sunspots. However, this cycle may well be characteristic of just the last 270 years or so: Around the team of this century, E.W. Maunder delved into old records. These revealed a surprising absence of sunspot activity for about 70 years, between 1645 and 1715. In this period, the sun's spectacular corona had vanished during total solar eclipses.

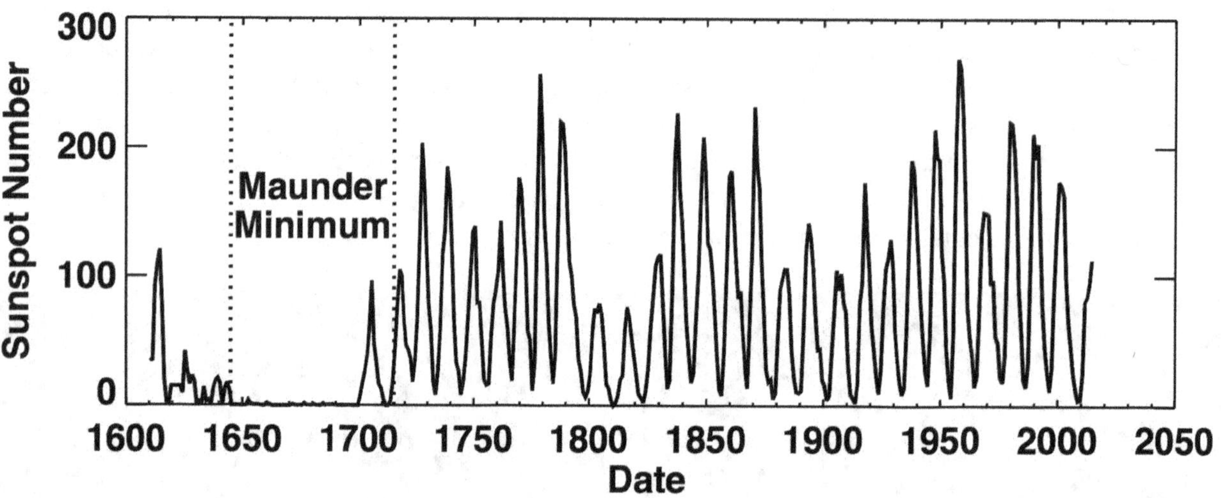

The decline of Sunspots during the Maunder Minimum (Source: The Solar Physics Group at NASA's Marshall Space Flight Center)

The breath-taking aurora displays in the higher latitudes had also become scarce. In fact, Edmund Halley of Halley's Comet fame, after a forty- year scan, detected the first aurora in 1716. And, suggestively, freak winters gripped England in the 1680s. The Thames river froze to a sheet of ice and even carnivals were organized on it!

Then came an uncanny twist. From nowhere near astronomy, though. Around 1920, A.E. Douglas analyzed tree-rings --- circular layers added to tree trunks annually. He discovered that the growth rate fluctuates periodically over one or two decades, like sunspots. Believe it or not, the growth of trees increases with sunspots: Initially Douglas didn't suspect the link. Later, he was astonished that this periodical variation of tree rings did not show up at all exactly during the seventy-year sunspot lull! mystery, if you want one.

Solar flares spotted in the rings of trees could help scientists pinpoint with more certitude the dates of ancient events. (National Geographic)

Recently, historical variations of sunspot activity have been detected from an unexpected foot-print left by solar winds. The radio-active isotope, Carbon 14, is present in the atmosphere in an almost constant amount, and shows up in the tree-rings. Though Carbon 14 keeps decaying off, cosmic rays colliding with atmospheric molecules, replenish the number. When sunspots are numerous, the associated solar winds of protons and other charged particles, sweep off the cosmic rays. So, the proportion of Carbon 14 dips. This can be detected in the tree-rings. There seem to have been a dozen researchable changes in the sunspot activity over the past few thousand years. Periods of high sunspot activity have been short and accompanied by warm weather in the higher latitudes. A boom in food production and population resulted. We may be in one such rosy epoch now -- but It is unlikely to last long. Then we confront the frightening prospect of mortifying winters, unthinkable food shortages and what not!

A line of current thinking connects sunspots and their cycles with the cyclical weather behavior patterns. The increased solar radiation associated with sunspots, heats the air and inflates it. This in fact, created friction and dragged down the American space craft, Skylab, prematurely in 1979. The air currents over the equator, which are speedier, owing to the earth's rotation, mingle with the cooler currents in the higher latitudes. This causes an eastward flow --- what is dubbed the subtropical jet stream. The farther north this stream reaches, the cooler the air it transports back to the mid-latitudes. In effect, the jet stream shifts. This could explain the colder winters in the eastern U.S. and the twenty-two-year drought cycle in the western U.S. Are the record winters of 1982, in many countries, due to dwindling sunspots too?

So, we are just about beginning to unravel the almost miraculous link between the fleeting sunspots and events on the earth, like the complex weather machinery. Don't be astonished if next, astronomers discover that a sunspots influence your metabolic processes! digestion or insomnia!

11

When the universe took a U-turn

As the 20th century was drawing to a close, astronomers had pretty much sewn up the universe It all started with the Big Bang. The so called Big Bang model arose from three main observations.

The first was the discovery in the 1920s that the Universe is expanding, in the sense that the basic constituents, the galaxies (as then believed) showed red shifts characteristic of receding objects. Furthermore as Hubble discovered, the farther the galaxy, the greater its speed of recession.

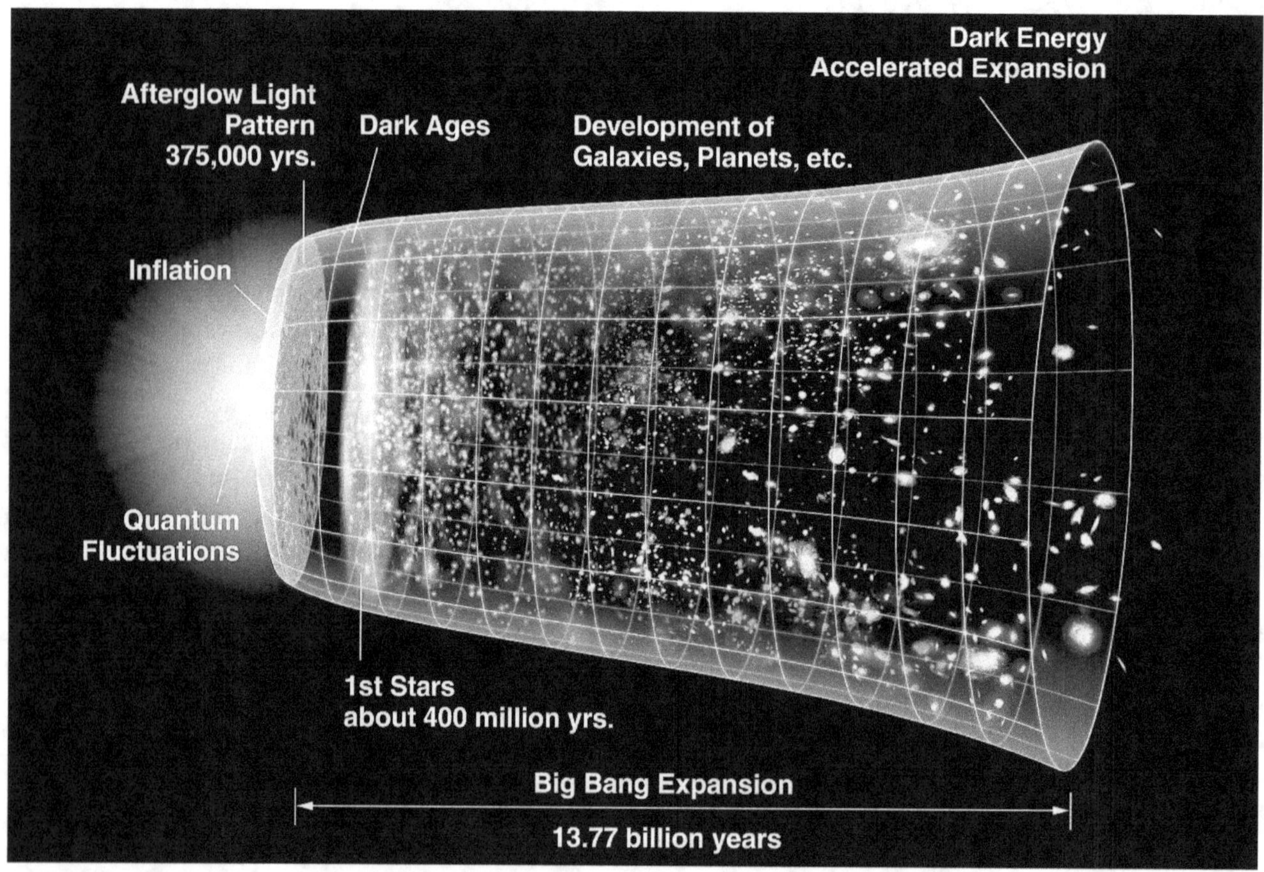

Timeline of the metric expansion of space, where space, including hypothetical non-observable portions of the universe, is represented at each time by the circular sections. On the left, the dramatic expansion occurs in the inflationary epoch; and at the center, the expansion accelerates (artist's concept; not to scale).NASA/WMAP Science Team - Original version: NASA;

Another important observation was about light element abundance - or overabundance - in the Universe. In the 1940s Gamow and coworkers provided an explanation for this. The early Universe must have been very hot and dense. The synthesis of light elements took place when the Universe was at a temperature of 19 billion K. However heavier elements were formed later, inside the stars, and were strewn about by supernova explosions.As if this job had been outsourscsed!

Finally there was a cosmic footprint of an explosion from a very early hot and dense state. This was the residual background radiation from that early event. In the present epoch. However the earlier intense radiation would have cooled, and it was calculated that it would be in the form of microwaves.Rather as in domestic appliances. Exactly such a cosmic background microwave radiation footprint was accidentally discovered in 1965 by Penzias and Wilson two radio astronomers They couldn't even guess what this radiation was.

The impressive Very Large Array in the heart of New Mexico

It was attributed to the pigeon droppings! This effectively overthrew a competing model of that time - the Steady State Model, which has now become history. So the picture to emerge was that the Universe was born in a titanic explosion or Big Bang, a name sarcaticaaly coined by Hoyle and made popular by Gamow. Exactly at the time of the Big Bang some fourteen billion years ago, it is reckoned, all the matter and energy of the Universe was concentrated at a single point just imagine! Here the density and curvature would be infinite. This is the Big Bang singularity. Following the Big Bang, matter and energy has been flung all round and even today the galaxies (or clusters of galaxies) are rushing outward due to that initial impact.

Galaxies rushing away due to the Big Bang (NASA)

The question that arises is, will the expansion of the Universe continue for ever, or would it slowdown to a halt and then re collapse? The answer to this would depend on the mass/energy density of the Universe. If this value is greater than a critical value, then the gravitational attraction will ultimately prevail over the expansion and the Universe would collapse. But if the density is less than the critical value, the Universe would go on expanding for ever. Observations seem to indicate that the density of the Universe was miraculously close to the critical value!

Further an observation of the speeds of rotation of the galaxies indicated that the galaxies themselves contained more matter than met the eye. This lead to ``Dark Matter'' being invoked. But what is dark matter? Dark matter has not been directly detected, nor can it be precisely characterized, even though there have been a number of possible guess candidates. For example invisible Black Holes or even difficult to detect brown dwarf stars. Exotic massive particles have also been thrown into the mix as also massive neutrinos or monopoles (that is only a north magnetic polefor examle)

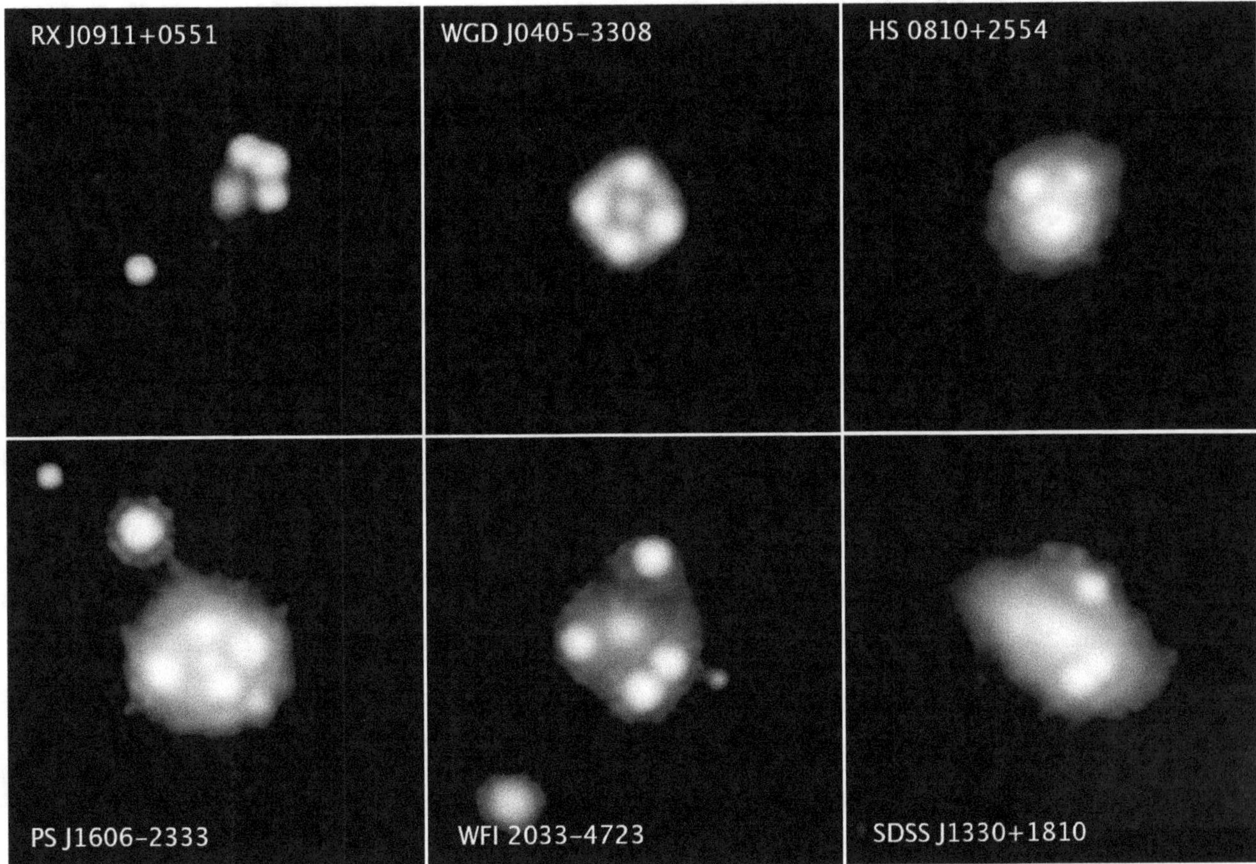

Hints of Dark Matter? (Hubble space telescope)

Then with dark matter thrown in, it was believed that the Universe had the critical density to reverse the expansion.

Though the Big Bang model could explain several observations, there were subtler questions which came to haunt. These were: How come the density of the Universe, which could have been anything, is in fact so miraculously close to the critical density in a process spread over billions of years? Is like winning a lottery ticket. More precisely such a close critical density today would imply that even after about a billionth of a second after the Big Bang the density was equal to the critical density accurate to some twenty five decimal places. Alternatively this means that the Universe or space is very flat. This need not have been so.

And then the Universe appears uniform on large scales. For instance the cosmic microwave background radiation is uniform in temperature to a high degree of accuracy.

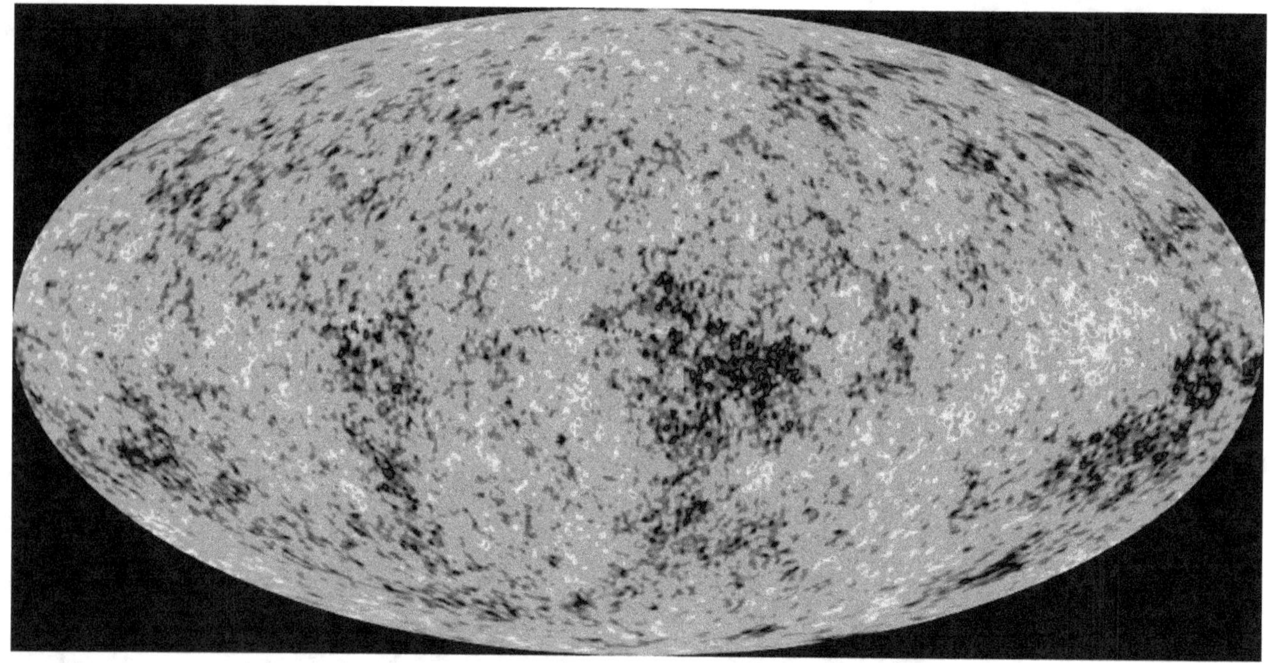

The full-sky image of the temperature fluctuations (shown as color differences) in the cosmic microwave background, made from nine years of WMAP observations. These are the seeds of galaxies, from a time when the universe was under 400,000 years old.Credits: NASA

How can this be so for regions separated by such vast distances. For since the Big Bang superflash itself there has not been enough time to connect distant places. This is called the horizon problem. Finally how do we account for the small scale inequalities or lumps in the Universe which we see as galaxies? In 1981 Alan Guth proposed his inflation Theory. According to this there was a super fast or super rapid expansion, forget the speed of light, in the early stages of the Universe, so that the size of the Universe exploded to several times its original size within a small fraction of a second.

To put it simply this super fast or exponential expansion flattens out the Universe, thus explaining the first problem. The horizon problem is also accounted for: Due to the super fast expansion or inflation, distant regions were much closer together than with a usual expansion. So they would be at the same temperature. Furthermore Quantum fluctuations would cause fluctuations in density, that is they would seed the formation of galaxies. Finally it may be added that given the inflationary scenario, the fact that exotic particles like magnetic monopoles are not detected is also explained. The rapid inflation would have diluted such particles and made them unobservable.

A time line of the Universe would be t = 10^{-43} secs that is the decimal point followed by 43 zeroes ! The Planck era of Quantum Gravity would have just ended and the Universe would be described by a Grand Unified Theoryin which all the forces would be indistisushable t = 10^{-35} secs, T = 10^{28} K. The Grand Unified symmetry is broken. And so on. The size of the Universe would still be only a millimeter across And so on WhenT ~ 10^9 K) Nucleosynthesis begins and nuclei of lighter elements like Helium and Lithium begin to form ectrons and nuclei combine to form neutral atoms as charged particles are no longer present. Interestingly Optical and Radio Astronomy cannot probe beyond this time t ~ 10^9 yrs, T ~ 10K. Finally Galaxy formation begins t ~ 10^{10} yrs when T ~ 2.7K. This is the Universe of today. The above was the model till 1997.

What happemed then ? That year, the author put forward an alternative model which in fact went against the then existing dogma. On the contrary, this model predicted a dark energy driven accelerating ever expanding and accelarating Universe. In 1998 dramatic confirmation for the new model came from the observations of Perlmutter, Schmidt, Riess and others. Clearly the limits of Standard Big Bang Cosmology had been reached in 1997. In his insightful book, The structure of scientific Revolutions, Thomas Kuhn points out that revolutions happen not increntally. But rather when the previous paradigm collapses. Starting with Kepler's laws centuries ago. Now it's a developing scene again, who knows with what twists and twirls?

12

A SHAKESPEAREAN TWIST -- THE TOPSY TURVY WONDER OF THE SOLAR SYSTEM

They trooped in one by one. Breathtaking Oberon, Titanic, Umbriel, Ariel and Miranda. All a chockful of surprises.

Not characters from a Shakespearean Comedy. But rather, bizarre frozen worlds in man's remote close up glimpse of the outer solar system They left astronomers gasping with awe and wonder. "... so totally different from anything we've seen! We're happily bewildered!" exclaimed one earth- based scientist. "Far more weird and wonderful than we ever expected .." marveled another. "It's all the strange places rolled into one", declared yet another NASA Scientist.

The story begins in 1738 in the family of a poor German Musician, Isaac Herschel. Friedrich Wilhelm was born that year and was trained to be a musician too. At the age of about fifteen, he started off as an oboist in the Hanover foot guards. When he was about twenty, Wilhelm emigrated to England on the advice of his father. Here he became an organist in the Chapel in the town of Bath. He taught music too and conducted a band for several years. Though some of his compositions are still played, rarely is this musician's name heard. However, it will echo in the world of stars, for a long long time. For, the poor music teacher blossomed into a renewed astronomer. Initially he could not afford a telescope. So, aided by his sister Caroline, who too sacrificed a promising musical career, he ground his own telescope lenses. And over the years, through sleepless nights he peered into the fathomless cosmos.

William Herschel (1738-1822) discovered Uranus, infrared radiation and moons of Saturn.

Between the 13th and 19th March 1781, Herschel prised out a very intriguing object in the Constellation of Gemini. A greenish disc that appeared to drift through the stars, ever so gently. What could it be?

Not a star. Even the nearest star is a mere pin point through the best telescopes. Never a disc.

Not a planet. For, there were only five planets up there. People could vouch for that for thousands of years. The giant astronomer of the 17th century, Johannes Kepler had in fact argued that there could be more than 5 planets in the sky.

At a meeting of the Royal Society, Herschel disclosed that he had discovered a comet.

It turned out to be quite uncomet like, though. It didn't display a tail. Nor was it fuzzy. And it lay well beyond Saturn, roving along a nearly circular orbit. A few months of close scrutiny forced Herschel to a reluctant conclusion. Incredible though it was, he had discovered a sixth and an altogether new planet.

Herschel christened the planet 'Georgium Sidus', after the then king of England, George III. But most astronomers preferred to call it 'Herschel'. However, within a few years, the name Uranus, suggested by the German Astronomer Johann Bode, outlasted the others. Uranus was the sky God of Greek Mythology, the father of Saturn and grandfather of Jupiter.

Uranus is uniquely tipped over among the planets in our Solar System. Uranus' moons and rings are also orientated this way, suggesting they formed during a cataclysmic impact which tipped it over early in its history. Credit: Lawrence Sromovsky, University of Wisconsin-Madison/W.W. Keck Observatory/NASA

Herschel wasn't the first to glimpse Uranus, though. It had been located about twenty times earlier: At least six times by the British astronomer Lloyd John Flamstead, almost a hundred years previously. With luck in fact Uranus can be just about visible to the naked eye. However, Herschel had hounded this object doggedly and was the first to realize that it was a planet. He even went on to unearth two of Uranus's satellites. More recently astronomers peering through telescopes discovered that Uranus displays faint bands a is Jupiter and Saturn, probably currents swirled up in a Methane rich atmosphere. This Hydrogen dominated planet is conjectured to have a rocky core at a temperature like 4000-degree C. A recent suggestion by an American Scientist is mind boggling! The high temperature and pressure in the planet could well have shattered the Methane into its constituents Carbon and Hydrogen and further, crushed the carbon itself into diamonds. In this unlikely event, Uranus and even Neptune may well launch a future, "GREAT SPACE DIAMOND RUSH!"

With 67 times the volume and 14.6 times the mass of the earth, Uranus is a giant planet. Being a little over 19 times as far off from the sun, it is a frozen world, the temperature swooping below minus 200-degree C, in its atmosphere. Hauling its family of 5 satellites, it swings round the sun once in 84 years. But unlike any other planet. It spins once every 17 hours in a plane that is almost at right angles to its orbit. So Uranus resembles a rotating wheel that rolls along the ground, while the other planets are more like upright spinning tops. There is a strange consequence. If you ever visit Uranus, and land on one of its poles for example, you will have a day that stretches for about 40 years, and an almost equally long night!

Uranus bounced back into the headlines just 4 years before its two hundredth 'Birthday'. On the night of the 10th March, 1977, Dr. J.C. Bhattacharya and his associates at the 40th telescope in Kavaloor, and two other American teams- one of them airborne- independently stumbled on an astonished discovery. While observing an eclipse of a star as it slid behind Uranus, they found that the star light flickered, like light passing through the chinks between quivering leaves. That is, they had detected a ring round Uranus, which consists of thousands of teeming ice like rocks, whirling round the planet separately. In fact, there turned out to be 5 separate rings, all within 40,000 to 50,000 kilometers of Uranus.

Early in 1986 the wonder space craft Voyager II swooped to within just 82,000 kilometers of the planet, for the first time in history, and in fact for a long long time to come. Obscure Uranus suddenly came alive as an enigmatic and astonishing world full of contradictions. And, it will take a few years to assess and analyze the flood of data relayed back.

As Voyager II plunged towards Uranus, its images beamed across 3000 million kilometers, revealed clouds swirling round the planet driven by high velocity, high altitude winds. Only, these gales blow in a direction opposite to that predicted by meteorologists! The atmosphere consists mostly of Hydrogen, Helium and Methane. Uranus glows with a bluish green hue because the Methane absorbs reddish light. At the same time, Voyager II rammed into a wall of radio disturbance which could only be generated by charged particles spiraling in a magnetic field surrounding the planet. And what a strange magnetic field it turned out to be! Voyager II discovered that this is about three times as powerful as the magnetic field surrounding the earth, and moreover coils round the planet, serpent like.

High-resolution views of Uranus' moons from Voyager 2 Voyager 2 captured the images that were used to create these mosaics of some of the moons of Uranus during its closest approach to the planet on January 24, 1986. NASA / JPL-Caltech / Ted Stryk

Voyager II clearly imaged for the first time, the rings of Uranus. In addition to nine known rings it disclosed two new rings and several ring fragments. As the Uranian rings came alive, scientists were surprised that they were crammed with rather large boulders, from a mmeter to a few kilometers across. Very unlike the rings of Saturn. Tinier dust particles hardly surfaced. An Infra-Red experiment threw up another enigmatic surprise! The pole turned towards the sun was actually slightly warmer than the winter pole, the one turned away from the sun!

Voyager II's expedition threw up ten new satellites of Uranus. But the satellites themselves provided the biggest thrills and upsets. If Uranus turned out to be a topsy- turvy world the means they were topsier and turvier! They are generally colorless, the surfaces being smeared with water ice and an as yet unidentified black carbonaceous material. Voyager II suggests that these moons are rockier and denser than these of Saturn, which is quite a paradox, because in the solar system, the farther away from the sun the object is, the less dense and recky it should be.

However, this puzzling finding would at least explain the totally unexpected degree of geological activity, the Uranian satellites exhibit.

Scientists gasped as the Voyager whizzed past each of the five known large Moons. Oberon the outer most, named after the Fairy King in Midsummer's Night Dream displayed a six-kilometer-tall mountain and craters splashed with a mysterious, dark coating. Probably some fluid had oozed out form the interior and solidified.

Titania- Shakespearean Oberon's Queen – next trouped into view, flaunting shining snowy patterns running by long cracks. Could some material have been vented up, which subsequently froze and crashed back on to the Satellite?

The darkest and oldest of the Uranian brerd Umbriel, taunted earth- based scientist by exhibiting a bright oval patch. Probably a crater, ever a hundred and fifty kilometers acres, which reflects almost a third of sunlight more than the rest of the surface.

Ariel zoomed in next, with breathtaking sinuous broad valleys, may be glacier tracks, and huge zig zagging canyons. Though there were few craters, grooves criss crossed all over the place and a huge crack crept over much of the moon. ".. a dramatic geology. .", exclaimed earth- based scientists.

If all this had tantalized scientists, the tiny innermost moon Miranda confounded and dumbfounded them! As Voyager II whooshed by, barely 43,000 kilometers away, it discovered an arabesque of deep grooves intersecting at steep angles craters, inexplicable streaks, cracks, a few kilometers deep trenches and to top it, cliffs soaring up to twenty kilometers. "It's amazing an object this small could have had that much internal activity!" one scientist exclaimed. Another declared, "If you took all the bizarre geological features and put them on one object, you'd het Miranda!" The enigma here is not merely the discovery of such a rocky object so deep in the solar system. The truth is that a small object like Miranda has no business to have been so geologically scarred. Only massive objects or those with an abnormally high content of rocky material can swizzle up enough internal heat for such geological activity. Given such a bizarre scenario, you can trust scientists to concoct a mindboggling theory! And they have. May be a comet crashed into and shattered the originally flat surfaced of Miranda. As the flung out chunks collapsed back, the internal rocks rained down on the top, making Miranda an inside out world!

Uranus has fully vindicated its discoverer's stubborn belief, "Whatever shines is worth observing", which is also the motto of Royal Astronomical Society. And more, the moral Uranus helds out is, the Universe never ceases to Startle!

13

THE EGO CENTRE OF THE GEO CENTRE

The Vatican

Rome, June 22, 1633. The gloomy wooden paneled hall of the Convent of Santa Maria Sopra Minerva, today a wing of the Post Office. Ten grim Cardinals, swathed in their swaying robes, confront a bearded, aging and visibly ailing scholar. He kneels down and in low tones chants from a document previously delivered to him:

"I, Galileo … swear that I have always believed … all that is held, preached and taught by the Holy Catholic and Apostolic church . . . I wrote and printed a book in which I discuss this new (sun- centered) doctrine already condemned and adduce arguments of great cogency in its favor … I abjure, curse, and detest the aforesaid errors and heresies … and I swear that in future I will never again say or assert…"

Galileo Galilei at the Inquisition. Credit: Getty Images

Thus terminated a drama that has been romanticized as the classic duel between reason and dogma, science and religion. The exact details have been a bit obscure. Old corpses are now being exhumed. After three hundred and fifty years. Actually, the little known truth is very surprising.

The origin of Galileo's crime lay in a book published ninety years earlier: "on the Revolution of the Heavenly Orbs", and dedicated, ironically, to people Paul III. In this treatise, a German Canon, Copernicus, inspired by a revolutionary and nearly two-thousand-year old concept, installed the sun at the center of the universe. The earth and other planets merely circumscribed it.

Apart from being incredible, the new model clashed with the earth or geo- centric cosmic blue- print, carefully fashioned by a series of Greek thinkers, terminating with Ptolemy. This model had dominated thinking for 1500 years and had even infiltrated into holy scriptures. The Bible, for example, proclaimed:

". . . the earth abideth forever. The sun also ariseth and the sun goeth down . . ."

To bypass controversy, Osiander, the publisher, had inserted without the dying Copernicus's knowledge, a conciliatory – and villainous – preface: The Copernican model was a mere alternative hypothesis. An inferior one, in fact.

By 1610, the Ptolemaic cosmos quaked. The telescope had just been accidentally invented.

Italian astronomer and physicist, Galileo Galilei (1564 – 1642) using a telescope, circa 1620.Hulton Archive/Getty Images

And a cantankerous Professor of Padua, Galileo Galilei, had swung one heaven wards. He beheld an astonishing spectacle. Four unsuspected 'stars' whirled round Jupiter. Not everything in the heavens circumscribed the earth. Next, and even astounding, the planet Venus displayed moon like phases. That meant, Venus did not orbit the earth. The Greeks had definitely erred about it. Contrary to popular belief, this did not demonstrate the earth's motion: The first physical proof came from a bewilderingly unexpected direction, as late as 1725.

Galileo, who relished publicity, announced some of his revolutionary discoveries in pioneering, popular language. Reaching out to the masses, he churned up quite a sensation. Soon thereafter, he visited Rome to be applauded by the Pope and other Jesuit scholars who even confirmed his startling findings.

By about 1612, Galileo's unwarranted tirades against contemporary scholars began surfacing. In some, he unjustly hogged credit for discoveries of sunspots and the like. Then, in 1613, at fifty, he first publicly defended Copernicus. Promptly, this evoked the admiration of several Jesuits, including Cardinal Barberini, the future Pope, who later extolled Galileo in verse.

It is surprising that, for the preceding twenty years, Galileo had actually refuted the sun centred Copernican blueprint, though he secretly couched it. Surprising because, the Church did not object to the Copernican theory aired as a pure hypothesis. In fact, Kingpin theologian, Cardinal Bellarmine declared: "... to say that the assumption that the Earth moves and the Sun stands still saves all the celestial appearances ... is to speak with excellent good sense ...

"... if there were a real proof that the sun is in the center of the universe ... we should rather have to say that we did not understand them (Passages from the scripture) than declare an opinion to be false which is proved to be true . . ."

Bellarmine, by the way, had earned immortal notoriety by being one of the Inquisitors who had condemned the radical, Giordano Bruno to the stake.

Around this time, a Jesuit admirer of Galileo, Father Castelli mouthed a few innocuous comments in private conversation. These did not really contradict Copernicus. Galileo as was now his wont, over- reacted. He issued two castigating letters of dubious merit. A copy of these documents reached the Holy Office with two deliberate distortions. His enemies made him sound downright outrageous. However, the Holy Office reacted graciously: Except that such strong terms[which weren't Galileo's anyway) should be abhorred, the content was innocuous enough. Galileo's snowballing band of enemies had been snubbed.

Then, about 1615, a book by a Jesuit priest, Rev. Fr. Paolo Antonio Foscarini swizzle up chaos. He upheld the Copernican model which, he argued, did not contradict Holy scriptures, and which could be suitably re-interpreted. This was Galileo's stance. He had even explained the Holy passages his own way. This was the church's stand too. The mood is reflected by a contemporary authority:" The intention of the Holy Spirit is to teach how to go to heaven and not how go the heavens."

The real clash lay elsewhere. The Church held, and rightly, that the Copernican model hadn't been irrefutably established --- yet. It was more an intellectual conviction.Next, if necessary. The Church itself would re- interpret Biblical passages. Not individuals. That could trigger off chaos. More importantly, Institutional Authority could be corroded. And all this, in the context of the Church's conflict with Lutherism. So, while Copernicus's book was merely suspended, pending a few trivial alterations. Father Antonio's treatise was totally banned and condemned.

In the process, Galileo was summoned by Bellarmine who subsequently delivered him a certificate. This cleared him but also communicated the Church's now hardening attitude: The Copernican doctrine should not be held or defended. All this purely incidentally, and cordially too. For soon the Pope himself granted Galileo a long audience.

But Galileo had tasted adulation and acclaim. He imagined that his personal prestige was at stake, ". . . In this quarrel, as if it were his own business . . .", a contemporary reported. So, in 1618, he privately circulated an incorrect theory that it is the earth's motion which raises tides. Next, he launched an uncalled for controversy with a Jesuit priest, Horatio Grassi of the college of Rome. The moot point: The unwitting Grassi and others had not credited Galileo – justifiably – with some discoveries. In a choleric passage Galileo avers: "You cannot help it . . .

that it was granted to me alone to discover all the new phenomena in the sky and nothing to anybody else... Galileo, with his superior intellect, not only rent Grassi and others apart, even where they were correct. He lampooned them at a personal level.

By now Galileo had perfected a style of not so subtle sarcasm, guaranteed to win him arguments and enemies. In 1623, when he was compiling his scathing critique on Grassi, old time admirer, Maffeo Barberini became the Pope. Promptly, Galileo dedicated the document to the new Pope Urban VIII, who in subsequent years granted him several audiences and favors.

But the new Pope was a nepotic, megalomaniac who exulted, "I know better than all the Cardinals put together. The sentence of a living Pope is worth more than all the decrees of a hundred dead ones." He openly admired Copernicus and declared, that the Church never "... would condemn his doctrine as heretical, but only as reckless."

From the various audiences with the Pope, Galileo inferred that he could air the Copernican model as a hypothesis. He even believed along with many others in the Holy office that the Pope would smuggle the new doctrine in through the back door. And he might well have. But something dramatic intervened;

In 1630, Galileo presented the Pope with a manuscript wherein three characters thresh out the merits of the Ptolemaic and Copernican systems. Salviati, the Copernican, sounds thoroughly convincing. He even cites Galileo's erroneous theory of tides to demonstrate the earth's motion. The pro Ptolemy Simplicio shows up as a thorough ass. Galileo even brands the Ptolemaic scholars as "Dumb idiots", Mental pygmies and so forth. The Pope didn't object to a discussion of the Copernican hypothesis. He flipped through the work and in fact, suggested its final title," Dialogue on the Great World Systems." Then he handed it over to his lieutenants for the usual screening. But Galileo, abetted by sympathizers in the Church manouvred to get the book published in 1652 without any real revision.

And Hell, or rather, the anti-Galileo lobby, which he himself had cultivated, went berserk. As Koestler in his book ,The Sleepwalkdrs underscores, "... the attitude of the Collegium Romanum and of the Jesuits in general changed from friendliness to hostility, not because of the Copernican views held by Galileo, but because of his personal attacks on leading authorities of the Order. They convinced the Pope that Simplicio was a caricature of himself. And unluckily,

Simplicio had, as a final desperate resort, blurted the Pope's pet dictum. In effect, to produce a phenomenon, God, starting at will, cannot be constrained to a single hypothesis, even if the said hypothesis plaints the phenomenon fully.

Actually Galileo hadn't intended at all to embarrass his mentor, but now the joke was on him. The furious Pope, his ego pricked, sacked his own secretary who had aided Galileo. According to a contemporary authority, he, "... treated the affair as a personal one." He blew it beyond proportions.

In retrospect, the immortalized trial was really an ironical farce: Neither defendant Galileo, nor the prosecution had a case! For, Galileo claimed - atrociously that the book actually refuted Copernicus! Nor had he contravened any stricture against himself. Nor was the Copernican theory officially declared heresy!

But then, the Pope was obsessed The inflated Galileo, who had, in his opinion deceived him, must be humiliated. " .. Matters cannot be let pass without some demonstration against his (Galileo's) person," he confided. So, Galileo was hit below the belt.

The Inquisition flaunted a document that has generated much controversy. Unsigned, without a seal or witness, it would be worthless in any court. This paper, allegedly issued to Galilee in 1816, forbade him from holding and defending the Copernican doctrine. But it went further and banned its discussion in any manner whatsoever."

That convicted Galileo. He crumbled mentally and recanted. Rather shamefully, in fact. Had Galileo, as extolled been the really inspired martyr of science, at least he would have retracted with greater dignity. Quite unnecessarily, he even offered to rewrite the Dialogues so as to vindicate Ptolemy.

The Church too had no conviction. So, contrary to popular belief, during the trial and the lifelong house arrest that was enacted, Galileo was treated; against all precedent, with utmost leniency. In fact, lavishly. The trial was a mockery of Justice. But all that mattered was, Galileo had been humbled.

The glorified cosmic duel was, in truth, a clash of two earthly egos, which sidelined the principles. The heliocentric model was irrelevant. As the head of the College of Rome lamented, if Galileo had not incurred the displeasure of the Company, he could have gone on writing freely about the motion of the earth to the end of his days."

The scandalous trial necessarily blemished the Church and made a martyr of Galileo, but it contributed to science unexpectedly! At last, the humiliated seventy- year- old Galileo abandoned petty polemics and harnessed his genius fruitfully. He produced his immortal treatise on mechanics.

Rome, May 1983, Exactly three hundred and fifty years after Galileo's confession, Pope John Paul II graciously concedes to a delegation of two hundred scientists that the Pope Urban VIII might have erred.

Thus climaxed years of Papal soul searching and re-thinking. Pope John Paul II has been underscoring that science and religion are not irreconcible. He has tried to symbolize this harmony by restoring Galileo's image. In 1980 he appointed a panel. of scholars to re-open the Galileo dossier. And, in July 1984, all surviving documents were released from the Vatican Archives.

Initial reactions: The heresy charge against Galileo, seem to have no foundation! and, "The judges who condemned Galileo committed an error."

More to follow . Infact already an orthodoxy has built up on topics like string theory which may perhaps go the same way.......